DIGITAL HISTOLOGY

FREDERICK W.K. KAN, Ph.D
Dept. of Anatomy and Cell Biology
Faculty of Health Sciences
Queen's University
Kingston, Ontario
Canada K7L 3N6

2004

Digital Histology

An Interactive CD Atlas with Review Text

Alice S. Pakurar, Ph.D.
Department of Anatomy and Neurobiology
Virginia Commonwealth University
Richmond, Virginia

John W. Bigbee, Ph.D.
Department of Anatomy and Neurobiology
Virginia Commonwealth University
Richmond, Virginia

A JOHN WILEY & SONS, INC., PUBLICATION

Published by John Wiley & Sons, Inc., Hoboken, New Jersey.
Published simultaneously in Canada.

Library of Congress Cataloging-in-Publication Data is available.

ISBN 0-471-64982-1

Printed in the United States of America.

10 9 8 7 6 5 4 3 2 1

CONTENTS OF THE CD

Contents of the Review Text

PREFACE

The concept for *Digital Histology* was born out of necessity. The project began as a series of digitized micrographs to be used in undergraduate histology courses at Virginia Commonwealth University. From there, it grew into a laboratory preview for the professional and graduate histology courses. However, before that transformation was complete, we encountered the reality of curriculum constraints and were faced with finding an alternative to traditional microscope laboratories. Initial versions of *Digital Histology* were used only for the basic tissue laboratories; however, it evolved very rapidly into its present form, which comprises a complete, computer-based histology laboratory.

The CD of *Digital Histology* consists of more than 700 light and electron microscopic images, each with superimposable labels and descriptive legends. The images are grouped into chapters and subchapters consistent with the traditional presentation of material in a histology course. An overview section, including original illustrations to clarify concepts, introduces each major topic. For review purposes, images can be viewed in random sequence, with or without labels and descriptions, and a quiz accompanies each major topic. *Digital Histology* also includes a "virtual slide" image, which can be zoomed in and out to simulate the use of a microscope.

Digital Histology is currently used both as a substitute for traditional microscope laboratories in the medical, dental, pharmacy, and dental hygiene curricula at Virginia Commonwealth University, and as an ancillary study aid for students in the graduate and physical therapy histology courses. We feel *Digital Histology* fills a unique niche. In this time of decreasing classroom hours in professional curricula, *Digital Histology* provides an interactive, annotated, digital atlas of histology images, insuring that each student can identify microscopic structures even without the aid of an instructor. Furthermore, it offers a realistic adjunct or alternative to histology laboratories at all levels of instruction, obviating the cost of microscopes and slide collections. The accompanying text, in outline format, follows the content of the CD and offers a concise reference for the student's review of key points.

Alice S. Pakurar, Ph.D.
John W. Bigbee, Ph.D.

ACKNOWLEDGMENTS

It is difficult to adequately thank all the people who contributed to *Digital Histology*.

Carole Christman, M.S., Ph.D., our medical illustrator, took our ideas for illustrations far beyond our limited vision and created works of art. She added immeasurably to *Digital Histology* in ways far beyond that of an illustrator. She challenged our knowledge and she kept us true to our craft. We are in her debt for her commitment, ability, insight, and knowledge.

Chris Stephens, Director of Educational Applications Development, VCU Office of Faculty Development, created the original template for *Digital Histology* and then kept adding to its intricacies. His knowledge, ability, and especially his extreme patience with two computer novices are gratefully appreciated.

John Priestly, M.A., instructional developer and "Defender of the English Language," designed the current menu for *Digital Histology* and added all the features that provide for the interactive ability of the program. His own ideas, initiatives, and knowledge contributed immeasurably to *Digital Histology*.

We would also like to thank Carol Hampton, M.M.S., Associate Dean for Faculty and Instructional Development, and Jeanne Schlesinger, M.Ed., Assistant Professor and Director of Instructional Development, for providing the resources of their offices during the development of *Digital Histology*. We especially appreciate Jeanne's overview of the project, from the cover design to the font size.

Finally, we are grateful to the Department of Anatomy and Neurobiology. John T. Povlishock, Ph.D., Chairman, generously supported our efforts both financially and emotionally. Special thanks go also to the faculty of the department, who provided images and graciously allowed us to use them in *Digital Histology*: Oliver Bögler, Ph.D., Shirley S. Craig, Ph.D., Jack L. Haar, Ph.D., George R. Leichnetz, Ph.D., Thomas M. Harris, Ph.D., Caroline Goode Jackson, Ph.D., Randall E. Merchant, Ph.D., Milton Sholley, Ph.D., David S. Simpson, Ph.D., and Andras K. Szakal, Ph.D. Thanks also to students Richard C. Zug and Judson Smith, who digitized and enhanced the quality of the images.

We would also like to thank each other for support, patience, balance and persistence!

CHAPTER

1

TISSUE PREPARATION AND MICROSCOPY

GENERAL CONSIDERATIONS

- Biological tissues must undergo a series of treatments to be observed with light and electron microscopes. The process begins by stabilization of the tissue with chemical fixatives. Next, the tissue is made rigid to allow sectioning. Finally, it is stained to provide contrast for visualization in the microscope.
- Steps in tissue preparation
    - Fixation
    - Dehydration
    - Infiltration and embedding
    - Sectioning
    - Staining

CHEMICAL FIXATION

- Preserves cellular structure and maintains the distribution of organelles.
- *Formaldehyde* and *glutaraldehyde* are the most commonly used chemical fixatives. They stabilize protein by forming cross-links between

Digital Histology: An Interactive CD Atlas with Review Text, by Alice S. Pakurar and John W. Bigbee
ISBN 0-471-64982-1

primary amino groups. Formaldehyde in solution is referred to as formalin.

➢ *Osmium tetraoxide* is a fixative used to preserve lipids, which aldehydes cannot do. Osmium combines with and stabilizes lipid and, in addition, also adds a brown color (light microscopy) or electron density (electron microscopy) at the site of the lipid. Osmium fixation is required for electron microscopy, especially to preserve the lipid in membranes.

DEHYDRATION, INFILTRATION, AND EMBEDDING

➢ Tissue water is not miscible with the embedding solutions and must be replaced using a series of alcohols at increasingly higher concentrations. This step is followed by alcohol replacement with an intermediate solvent that is miscible with both alcohol and the embedding solutions.

➢ Infiltration and embedding. The liquid form of the embedding compound, for example, paraffin wax or epoxy plastic, replaces the intermediate solvent. The liquid embedding medium is allowed to solidify, thereby providing rigidity to the tissue for sectioning.

SECTIONING

➢ The embedded tissue is cut thin enough to allow a beam of light or electrons to pass through.

➢ Section thickness
- Light microscopy. 1–20 microns
- Electron microscopy. 60–100 nanometers

➢ Section planes
- Cross-section (cs) or tranverse section (ts) is a section that passed perpendicular to the long axis of a structure.
- Longitudinal section (ls) is a section that passed parallel to the long axis of a structure.
- Oblique (tangential) section is any section other than a cross- or longitudinal section.

STAINING

➢ Most tissues have no inherent contrast; thus, stains must be applied to visualize structures.

➢ Conventional staining. Relies mostly on charge interactions.

- Light microscopy
  - *Hematoxylin and eosin (H&E).* These two dyes are the most commonly used stains in routine histology and pathology slides. Most conventional stains bind to tissue elements based on charge interactions, that is, positive charge attraction for a negatively charged structure. Hematoxylin binds to negatively charged components of tissue, the most prominent being nucleic acids. Hematoxylin imparts a purple/blue color to structures and, therefore, the nucleus and accumulations of rough endoplasmic reticulum in the cytoplasm, which contains large amounts of nucleic acid, appear blue or purple in sections.
  - Structures, like the nucleus and rough endoplasmic reticulum that stain with hematoxylin, are referred to as *basophilic* or "base loving." The term *basophilia,* refers to the property of a structure or region that stains with a basic dye, such as hematoxylin. Structures that stain with eosin, for example, the cytoplasm of most cells and collagen fibers, appear pink or orange and are referred to as *eosinophilic.*
- Electron microscopy
  - Images in the electron microscope are produced by passing a beam of electrons though the tissue that has been "stained" with salts of heavy metals, usually lead (*lead citrate*) and uranium (*uranyl acetate*). These metals bind to areas of negative charge and block the passage of the electrons through the section, resulting in a dark area in the electron micrograph. Electron density is also achieved using osmium tetroxide, which also serves as a lipid fixative.
  - Areas or structures in tissue that bind the metals are referred to as *electron dense.* Areas where the metals do not bind appear light and are referred to *electron lucent.*

➢ Histochemical staining. Localizes chemical groups
- *Osmium tetroxide.* Stains lipids
- *Periodic acid–Schiff stain (PAS).* Stains carbohydrates

➢ Immunocytochemistry. Localization of specific antigens in cells using labeled antibodies

➢ *In situ* hybridization. Detection of messenger RNA or genomic DNA sequences using labeled nucleotide probes

ARTIFACT

➢ The term artifact is used to refer to any feature of a tissue section that is present as a result of the tissue processing. These include tears and folds, shrinkage, spaces resulting from extracted cellular contents (e.g., lipid, precipitates), and redistributed organelles.

MICROSCOPY

- Properties
  - Resolution is the smallest degree of separation at which two objects can still be distinguished as separate objects and is based on the wavelength of the illumination.
    - Light microscopy. Approximately 200 nm
    - Electron microscopy. Approximately 1 nm
  - Magnification. Enlargement of the image
- Bright field microscope
  - An image is formed by passing a beam of light through the specimen and then focusing the beam using glass lenses.
  - The bright field microscope is called a compound microscope because it uses two lenses, objective and ocular, to form and magnify the image. The compound microscope typically has a total magnification range of 40–1000 times.
- Electron microscope
  - Transmission electron microscope (TEM)
    - An image is formed by passing a beam of electrons through the specimen and focusing the beam using electromagnetic lenses.
    - Similar arrangement of lenses is used as with optical microscopy; magnification is up to 400,000 times, which is sufficient to visualize macromolecules (e.g., antibodies and DNA).
  - Scanning electron microscope (SEM). The image is formed by electrons that are reflected off the surface of a specimen, providing a three-dimensional image; magnification ranges from 1–1000 times.
- Freeze fracture technique
  - This technique is used to examine the number, size, and distribution of membrane proteins.
  - A tissue is frozen and mechanically fractured; the exposed membrane surface is coated with a thin metal film called a "replica."
  - The replica is viewed by TEM. Membrane proteins appear either as bumps or pits in the replica.

Section Interpretation

- In histology, three-dimensional tissues are viewed in two dimensions; therefore, it is extremely important to learn to visualize the three-dimensional structure from the two-dimensional image. For example, a cross-section through a tubular structure appears as a ring, whereas a longitudinal section appears as two parallel bands. As an added challenge, most sections pass obliquely to these perpendicular axes and, thus, require further "mental gymnastics."

Units of Measure

- Millimeter (mm) = 1/1000 meter, 10^{-3} M
- Micron, micrometer (μm) = 1/1000 mm, 10^{-3} mm, 10^{-6} M
- Nanometer (nm) = 1/1000 μm, 10^{-3} μm, 10^{-9} M
- Ångström unit (Å) = 1/10 nm, 10^{-10} M

CHAPTER

2

Cell Structure

General Concepts

➢ Heirarchy of body organization
  - Cells
  - Tissues (epithelium, muscle, connective, nervous)
  - Organs (stomach, heart, skin, lung)
  - Organ systems (digestive, respiratory, excretory)
  - Individual

➢ Although there are approximately 200 different cell types in the body, cells are more alike than different. Specialization of function, for example, glandular cells for secretion or muscle cells for contraction, is really an emphasis of a function that all cells possess to some degree. In some cases, cells have become so specialized that some functions are lost altogether (e.g., cell proliferation).

➢ Cells are the structural units of all living organisms.

➢ Cells vary in size and shape according to location and function.

➢ Cells vary in internal structure depending upon their function.

➢ Cells vary in their life history, for example, rates of cell renewal.

➢ Major compartments of the cell
  - Cytoplasm. Provides an aqueous matrix containing the internal

Digital Histology: An Interactive CD Atlas with Review Text, by Alice S. Pakurar and John W. Bigbee
ISBN 0-471-64982-1

structures of the cell and allowing for the cytosolic metabolic pathways (e.g., glycolysis).

- Nucleus

CELL MEMBRANES

➢ All membranes have a similar structure and are referred to as *unit membranes*. A specialized unit membrane forms the surface boundary of the cell and is called the *plasma membrane*. Other membranes are present within the cells, where they form mitochondria, endoplasmic reticulum, or nuclear envelope, for example. The structure of the unit membrane cannot be resolved with the light microscope; however, at high magnification with the electron microscope, it appears as a trilaminar dark-light-dark band.

- *Fluid mosaic model of membrane structure*
    - ◆ *Phospholipid bilayer* consists of two leaflets of phospholipids.
        - ■ The polar, phosphate head groups face the surface of the membrane.
        - ■ The hydrocarbon tails form the hydrophobic core of the membrane.
    - ◆ Membrane proteins
        - ■ *Integral membrane proteins* are proteins that extend into one or both of the phospholipid layers. Proteins that extend across both of the phospholipid layers are called transmembrane proteins.
        - ■ *Peripheral membrane proteins* are either associated with the polar head groups of the phospholipids or with integral membrane proteins. They do not contact the hydrophobic core of the membrane.
    - ◆ *Glycocalyx* is composed of complex carbohydrates lying on the external surface of the plasma membrane. The carbohydrates are covalently attached to proteins or lipids.

➢ Functions of the plasma membrane

- Membrane transport
    - ◆ *Diffusion*
        - ■ Passive diffusion
        - ■ Facilitated diffusion. Utilizes transmembrane proteins to increase the permeability of the membrane to certain materials.

- ◆ *Active transport*. Energy-requiring process of moving materials across the membrane
- ◆ *Vesicular transport*
  - ■ *Endocytosis*. Internalization of small membrane vesicles formed from the plasma membrane
    - – *Pinocytosis* ("cell drinking"). Uptake of fluid into the cell by a continuous process
    - – *Receptor-mediated endocytosis*. Requires receptor-ligand binding for vesicle formation and internalization
  - ■ *Phagocytosis* ("cell eating"). Ingestion of large particles (e.g., bacteria) into the cell; prominent in macrophages and leukocytes
  - ■ *Exocytosis*. Fusion of cytoplasmic vesicles with the plasma membrane and release of the vesicle contents to the outside of the cell
    - – *Constitutive exocytosis*. Continuous process (e.g., renewal of plasma membrane)
    - – *Regulated exocytosis*. Requires an extracellular signal for vesicle fusion and release (e.g., hormone secretion)
  - ■ *Transcytosis*. Uptake of material on one side of a cell followed by transport and release from the opposite surface
- ● Cell adhesion. Proteins provide cell-to-cell attachment and cell-to-extracellular matrix anchorage.
- ● Intercellular communication and signal transduction. Transmembrane proteins assemble to form pores between cells. Transmembrane, receptor proteins initiate intracellular signaling pathways following interaction with extracellular signals.

NUCLEUS

- ➢ Houses the DNA; produces ribosomes and messenger RNA
- ➢ Components
  - ● *Nuclear envelope*
    - ◆ Composed of two unit membranes, *inner* and *outer nuclear membranes*, which are separated by the *perinuclear space*; outer membranes and space are continuous with those of the endoplasmic reticulum.
    - ◆ Outer membrane possesses ribosomes.
    - ◆ *Nuclear pores*. Perforations in the nuclear envelope, provide

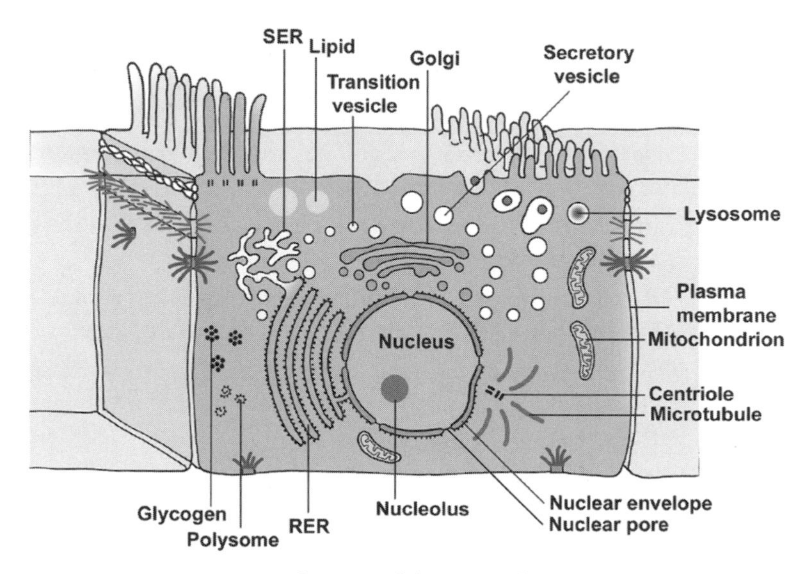

FIGURE 2.1. Structural features of a typical cell.

direct, bidirectional continuity between the contents of the nucleus and the cytoplasm.

- Inner and outer nuclear membranes become continuous at the rim of the pore.
- Pores are surrounded by an octet of proteins with a central granule comprising the *nuclear pore complex.*

◆ *Nuclear lamina.* Intermediate filaments on the inner nuclear membrane provide support for the nuclear envelope.

- Nucleolus
  - Site of ribosomal RNA (rRNA) synthesis and initial ribosome subunit assembly
  - Subdivisions of the nucleolus
    - *Nucleolar organizing centers.* Pale staining regions containing DNA sequences that encode rRNA
    - *Pars fibrosa.* Electron dense fibrillar region composed of rRNA transcripts
    - *Pars granulosa.* Granular-appearing region composed of maturing ribosome particles
- Chromatin
  - DNA plus protein, mostly histone protein

- ◆ Chromatin exists in transcriptionally active and inactive states.
  - ■ *Euchromatin.* Refers to the state of chromatin that is transcriptionally active, dispersed, and pale staining
  - ■ *Heterochromatin.* Refers to the state of chromatin that is transcriptionally inactive, condensed, and dark staining
- ● Nucleoplasm. Similar to cytoplasm, an aqueous matrix with cytoskeletal elements

ENDOPLASMIC RETICULUM

- ➢ Intracellular system of membranes
- ➢ *Rough endoplasmic reticulum (RER)*
  - ● Flattened membrane sacs; can occur singly or as multiple, parallel stacks
  - ● Continuous with the nuclear envelope
  - ● Possesses ribosomes on the cytoplasmic surface
  - ● Site of protein synthesis and some phospholipid synthesis
- ➢ *Smooth endoplasmic reticulum (SER)*
  - ● Tubular membranous structures in a meshwork configuration that is continuous with rough endoplasmic reticulum; lack ribosomes
  - ● Highly specialized in muscle cells where it is called the *sarcoplasmic reticulum*
  - ● Functions
    - ◆ Synthesis of triglycerides, cholesterol, and steroid hormones
    - ◆ Detoxifies drugs
    - ◆ Stores and mobilizes calcium

RIBOSOMES

- ➢ *Ribsomes* are composed of two subunits containing rRNA and proteins.
- ➢ Translation of messenger RNA (mRNA) to produce protein
- ➢ Distribution
  - ● Free in the cytoplasm. *Polysomes*, spiral clusters of ribosomes along a mRNA molecule; synthesize proteins for use in the cytoplasm, mitochondria, peroxisomes, and nucleus
  - ● Associated with membranes
    - ◆ Attached to the endoplasmic reticulum or outer nuclear membrane

- ◆ Synthesize: (1) proteins for incorporation into secretory vesicles for release outside the cell; (2) hydrolytic enzymes in lysosomes; (3) integral membrane proteins; and (4) proteins that function in the endoplasmic reticulum and Golgi apparatus

GOLGI APPARATUS

- ➢ The *Golgi apparatus* is composed of flattened, membranous sacs (*Golgi cisterns*), usually located near the nucleus. Has no structural continuity with the endoplasmic reticulum
- ➢ Modifies proteins formed in the RER (post-translational modification), for example, glycosylation and phosphorylation, and packages them into vesicles
- ➢ Components
  - *Transition/transfer vesicles* deliver proteins synthesized in the rough endoplasmic reticulum.
  - Transition vesicles fuse with the *cis* or *forming face* (convex surface) of the first Golgi cistern.
  - Proteins move between cisterns in vesicles formed at the margins of the cisterns.
  - *Trans* or *maturing face* (concave surface), represented by the last Golgi cistern, is the site of final vesicle formation.
  - *Trans Golgi network* is the collection of newly formed vesicles that are then routed throughout the cell.
- ➢ Fates of Golgi vesicles from the *trans* Golgi network
  - Fuse with plasma membrane, thereby supplying new lipids and proteins to the membrane
  - Form Golgi hydrolase vesicles ("pre-lysosomes")
  - Form secretory vesicles
  - Return to the endoplasmic reticulum or Golgi

LYSOSOMES

- ➢ *Lysosomes* are membrane-bound vesicles that serve as sites of intracellular digestion.
- ➢ Abundant in cells with high phagocytic activity (macrophages, neutrophils)
- ➢ Formation and function
  - Formed at the *trans* Golgi network; contain multiple (>40) inactive hydrolytic enzymes (e.g., protease, lipase, nuclease)

- Fuse with endosomes, phagosomes, or autophagosomes, which is followed by a decrease in the luminal pH that activates the hydrolases
- Hydrolases degrade the contents of the lysosome; undigestible materials are retained as residual bodies.

SECRETORY VESICLES

➢ *Secretory vesicles* constitute those vesicles derived from the *trans* Golgi network that are either not retained in the Golgi or endoplasmic reticulum or not destined for lysosomes.

➢ Contain highly concentrated secretory product in single membrane-enclosed structures

➢ Transported to the cell surface and fuse with the plasma membrane either in a continuous, constitutive mode or in a regulated fashion that requires an external signal

➢ *Zymogen granules.* Secretory vesicles that contain enzymes

MITOCHONDRIA

➢ *Mitochondria* (singular, mitochondrion). Sites of ATP production in the cell

➢ Spherical to ovoid shape, 1–10 μm; may be dispersed throughout the cytoplasm or clustered (e.g., at the poles of the nucleus)

➢ Composed of *inner* and *outer unit membranes*. The inner membrane is highly folded to form *cristae*, which provide increased surface area.

➢ *Intercristal space.* Located between the cristaes and is occupied by *matrix*

➢ Contain the enzymes for ATP production. Krebs cycle enzymes are located in the matrix whereas those for the electron transport system are located in the inner membrane.

➢ Matrix contains mitochondrial DNA, RNA, and electron-dense, *calcium-containing granules.*

➢ Mitochondria are self-replicating.

PEROXISOMES

➢ *Peroxisomes.* Membrane-bound vesicles containing oxidative enzymes (e.g., catalase and beta-oxidation enzymes)

➢ Carry out fatty acid oxidation and detoxification of alcohol

LIPID DROPLETS

- *Lipid droplets* consist of accumulations of cholesterol and triglycerides and are not surrounded by a membrane.
- Can occur as numerous small accumulations or as a single large droplet, as in adipose cells

GLYCOGEN GRANULES

- *Glycogen granules.* Occur in small clusters and appear highly electron dense; not surrounded by a membrane; stain magenta with the periodic acid-Schiff (PAS) reaction for carbohydrate
- Storage form of glucose
- Present in all tissues but highest in liver and striated muscle

PIGMENT

- *Lipofuscin pigment*
  - Residue of indigestable material that is contained within membrane-bound vesicles
  - Derived from late-stage lysosomes
- *Melanin*
  - Primarily synthesized by *melanocytes* but can occur in other cell types
  - *Melanosome.* Membrane-bound vesicle in the cytoplasm containing the melanin pigment
  - Main pigment responsible for hair and skin color

CYTOSKELETON

- Gives shape and support for the cell, provides cell motility, and facilitates intracellular transport
- Classification
  - *Microfilaments.* 4–6 nm filaments composed of actin; function in cell movement and extension of cellular processes
  - *Intermediate filaments.* Structurally and chemically heterogeneous; 8–10 nm filaments that are cell-type specific; function in structural support
  - *Microtubules.* 18–20 nm tubules composed of α and β tubulin; multiple functions within the cell. Microtubules form a number of structures in the cell including:

- *Centriole*. A short rod-like structure composed of nine sets of three microtubules; centrioles occur in pairs near the nucleus and are oriented at right angles to each other. The pair is called a *diplosome* and the region of the cytoplasm where they are located is the *centrosome* or the *microtubule organizing center*.
- *Basal body*. Structurally similar to centrioles; located at the base of cilia and flagella, providing support and serving as the source of the microtubule core of these structures
- *Axoneme*. Forms the core of cilia and flagella and provides for the movement of these structures; the axoneme consists of nine pairs of microtubules surrounding two central unpaired microtubules. These microtubules are connected to the basal body.
- *Mitotic spindle*. Individual microtubules that extend from the centrioles to the kinetochore of chromatids during cell division

CELL CYCLE

- The cell cycle is the period that extends from the time a cell comes into existence, as a result of cell division, until it completes its own cell division.
- Phases of the cell cycle
  - *Interphase*
    - *G_0 Phase*. Special G_1 phase for quiescent (nondividing) cells
    - *G_1 Phase*. Initial period of cell growth
    - *S Phase*. Period of DNA synthesis, chromosome duplication, and duplication of the centrioles
    - *G_2 Phase*. Preparation for cell division cell division, leads to M phase (mitosis)
  - *Mitosis (M phase)*
    - *Prophase*
      - Chromatin condenses.
      - Nuclear envelope disassembles.
      - Nucleolus disappears.
      - Centrioles migrate to opposite poles of the nucleus.
    - *Metaphase*
      - *Mitotic spindle* forms.
      - *Metaphase plate* forms.
    - *Anaphase*
      - Sister chromatids separate.

- Chromatids begin migrating to opposite poles of the cell.
- Cleavage furrow form.

◆ *Telophase*

- Chromatids complete segregation to opposite spindle poles.
- Chromatid DNA disperses.
- Nucleolus and nuclear envelope reform.
- Contractile ring forms, leading to *cytokinesis* and separation of independent daughter cells.

➢ *Meiosis*

- Cell division process restricted to the germinal cells in the gonads (ovary and testis) for the production of gametes (ovum and sperm), respectively.
- Results in reduction of the chromosome number by one-half (haploid); fusion of the gametes to form a zygote reconstitutes the normal diploid number of chromosomes present in somatic cells.
- Results in four daughter cells
- Meiosis occurs in two stages:

◆ *Meiosis I (reductional division)*

- At anaphase, duplicated chromosomes of each homologous pair migrate to the opposite spindle pole.
- Each daughter cell contains one-half the number of chromosomes. Each chromosome still consists of two sister chromatids.

◆ *Meiosis II (equational division)*

- Begins soon after meiosis I
- Same phases as mitosis but involves segregation of sister chromatids to daughter cells by the same phases seen during mitosis

STRUCTURES IDENTIFIED IN THIS SECTION

Staining
- Basophilic
- Electron dense
- Electron lucent
- Eosinophilic
- Osmium
- Periodic acid-Schiff (PAS)

Cell shapes
- Columnar
- Cuboidal
- Spherical
- Spindle
- Squamous
- Stellate

Centrioles

Centrosome

Diplosome

Glycogen granules

Golgi apparatus
- *cis* (forming) face
- Negative Golgi
- *trans* (maturing) face
- Vesicles

Lipid

Lipid droplets

Lysosome

Mitochondria
- Cristae
- Matrix
- Matrix granules

Mitosis
- Anaphase
- Centriole
- Cytokinesis
- Interphase
- Metaphase
- Metaphase plate
- Microtubules
- Prophase
- Spindle fibers
- Telophase

Nucleolus
- Nucleolar organizing center
- Pars fibrosa
- Pars granulosa

Nucleus
- Euchromatic nucleus
- Euchromatin
- Heterochromatic nucleus
- Heterochromatin
- Nuclear envelope
  - Inner nuclear membrane
  - Nuclear pores
  - Outer nuclear membrane
  - Perinuclear space (cistern)

Plasma membrane
- Intramembranous proteins (particles)

Polysomes

Ribosomes

Rough endoplasmic reticulum (RER)
- Nissl bodies (Nissl substance)

Secretory vesicles

CHAPTER

3

EPITHELIAL TISSUES

GENERAL CONSIDERATIONS

- ➢ Classification of epithelial tissues
  - *Lining and covering epithelia*
    - ◆ Form the boundary between external environment and body tissues
      - ■ Cover body surfaces (e.g., the epidermis of the skin) and lines the lumens of internal organs that open to the exterior of the body
      - ■ Line body cavities (e.g., peritoneal cavity) and covers the exterior of organs that project into these cavities
      - ■ Line blood and lymph vessels
    - ◆ Cell shape and number of layers correlate with the function of the epithelium.
  - *Glandular (secretory) epithelia*
    - ◆ Develop from a lining or covering epithelium by invagination into the underlying connective tissue
    - ◆ Form exocrine and endocrine glands (See below.)
- ➢ General features of all epithelial tissues
  - Highly cellular (sparse intercellular space)
  - Numerous intercellular junctions for attachment and anchorage

Digital Histology: An Interactive CD Atlas with Review Text, by Alice S. Pakurar and John W. Bigbee
ISBN 0-471-64982-1

- Avascular
- High regenerative capacity, especially in epithelial membranes, to replace continual sloughing of cells from free surface
- Most rest on a *basement membrane.*
    - The basement membrane is composed of a *basal lamina* and a *reticular lamina*.
        - The basal lamina is secreted by the epithelial cells and consists of the *lamina lucida* and the *lamina densa*. A similar structure is also present in muscle and nervous tissue, where it is referred to as an *external lamina*.
        - The reticular lamina is secreted by fibroblasts located in the underlying connective tissue.
    - Functions of the basement membrane
        - Provides support and attachment for the epithelial cells
        - Selective diffusion barrier
- Free and basal surfaces
    - *Basal surface* contacts the basal lamina of the basement membrane.
    - *Free surface* interfaces with the external environment or spaces within the body.
    - *Polarity*. A polarized cell is one that exhibits contrasting properties or structures on opposite sides of the cell. Because epithelial tissues face a free surface, the function of the apical surface is often very different from that at the base of the cell. This diversification is reflected by the nonhomogeneous distribution of organelles.

LINING AND COVERING EPITHELIAL TISSUES

METHOD OF CLASSIFICATION

➢ Classification by number of layers

- *Simple epithelium*
    - One cell layer thick
    - All cells rest on the basement membrane (basal surface) and all cells face the free surface.
- *Stratified epithelium*
    - More than one cell layer thick

- ◆ Only the deepest layer of cells contact the basement membrane and only the superficial-most cells have a free surface.
- ◆ Named according to the shape of the cells at the free surface omit.

➢ Classification by shape of surface cells

- *Squamous*
  - ◆ Cells are much wider than tall, resembling a "fried egg."
  - ◆ Nucleus is highly flattened.
- *Cuboidal*
  - ◆ Cells are of equal height and width.
  - ◆ Nucleus is spherical.
- *Columnar*
  - ◆ Cells are much taller than they are wide.
  - ◆ Nucleus is oval shaped, generally located toward the base of the cell.

TYPES OF LINING AND COVERING EPITHELIUM

➢ Simple epithelial tissues

- *Simple squamous*
  - ◆ Allows for rapid diffusion across the epithelium

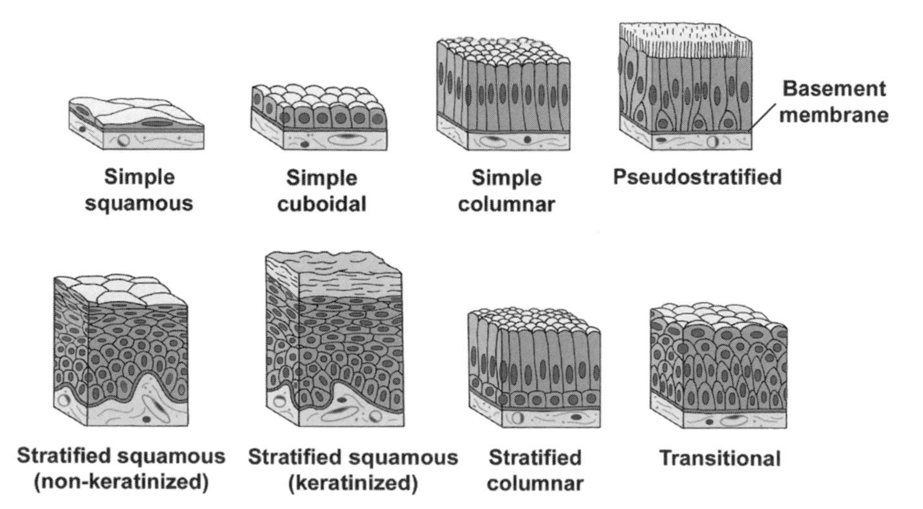

FIGURE 3.1. Types of lining and covering epithelia.

- Forms the lining of blood vessels, alveoli of the lungs, and internal body cavities
- *Simple cuboidal*
  - Lines and absorbs
  - Forms the walls of ducts and tubules
- *Simple columnar*
  - Lines and absorbs
  - Forms the lining of the intestines and gall bladder
- *Pseudostratified*
  - Cells are of various heights. All cells rest on the basement membrane, but only the tallest cells reach the free surface. Variation in height of the cells and the location of nuclei give the appearance of a stratified epithelium. Frequently ciliated.
  - Provides protection
  - Forms the lining of much of the respiratory tract and much of the male reproductive system

➢ Stratified epithelial tissues

- *Stratified squamous*
  - Protects from physical abrasion and prevents desiccation
  - Types
    - *Nonkeratinized (moist).* Lining of wet cavities, including the mouth, esophagus, rectum, and anal canal; surface cells are nucleated and living.
    - *Keratinized (dry).* Epidermis of the skin; surface cells are nonliving.
- *Stratified cuboidal/columnar.* Lines the larger ducts of exocrine glands.
- *Transitional*
  - Protective function; constructed to expand with distension of the hollow organs it lines
  - Unique to the urinary system; lines the urinary bladder and ureter

SURFACE SPECIALIZATIONS

➢ *Microvilli*

- Finger-like extensions from the free surface of the cell, about 1 micron in height

- Are present in large numbers on each cell and, collectively, are called a *brush* or *striated border*
- Contain a core of actin microfilaments
- Are relatively nonmotile
- Increase surface area for absorption
- Prominent on cells lining the digestive tract and proximal tubules in the kidney

➢ *Stereocilia*

- Large, nonmotile microvilli; not cilia
- Contain a core of actin microfilaments
- Increase surface area
- Present on cells lining the epididymis and ductus deferens in the male reproductive tract

➢ *Cilia*

- Multiple hair-like extensions from free surface of the cell; 7–10 microns in height
- Highly motile; beat in a wave-like motion
- Function to propel material along the surface of the epithelium (e.g., in the respiratory system and the oviduct of the female reproductive system)
- Core of a cilium is called the *axoneme,* in which nine pairs of microtubules surround a central pair of microtubules (9 + 2 arrangement).
- The axomene of each cilum originates from a *basal body* that is located at the apex of the cell and is composed of nine triplets of microtubules.

CELL JUNCTIONS

➢ Specialized structures of the plasma membrane that:

- Attach and anchor cells
- Establish apical and basolateral membrane domains by sealing adjacent plasma membranes
- Provide channels for ionic and metabolic coupling

➢ Not restricted to epithelial cells; cell junctions occur, however, in large number in epithelial tissues to resist the physical forces acting on the cells.

➢ Types

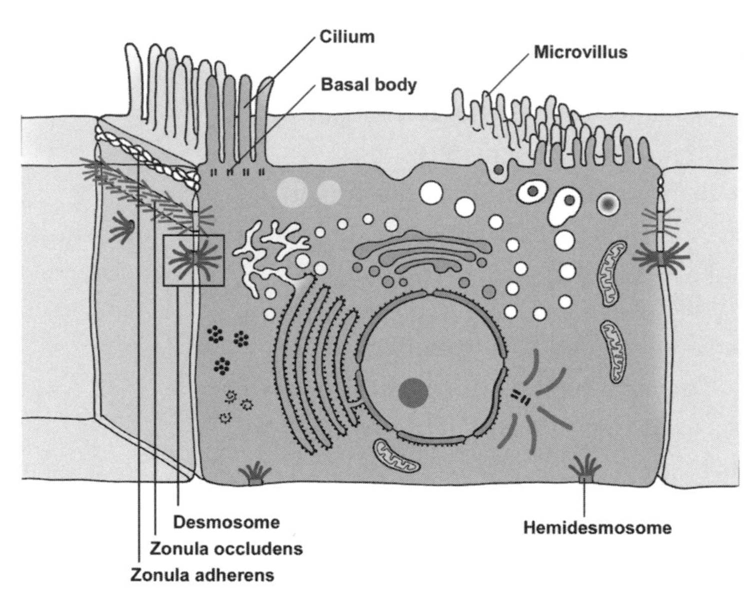

FIGURE 3.2. Cell junctions and surface specializations.

- *Tight junction (zonula occludens)*
  - Belt-like, barrier junction around apex of the cell
  - Provides close apposition of adjacent plasma membranes and occludes the intercellular space
  - Functions
    - Prevents diffusion of material between the intercellular space and the lumen of the organ
    - Establishes apical and basolateral membrane domains in the cell by preventing the lateral migration of proteins in plasma membrane
- *Adherent junctions*
  - Attach cells to each other and anchor them to the basal lamina; no fusion of the plasma membrane
  - Types of adherent junctions
    - *Belt desmosome (zonula adherens).* Belt-like junction that encircles the apex of the cell like a barrel strap and is located immediately beneath the zonula occludens; serves to attach adjacent cells together; associated with actin filaments.
    - *Spot desmosome (macula adherens).* Disk-like junctions scattered over the surface of the cell, which are paired with similar structures in adjacent cells; associated with intermediate filaments (e.g., keratin filaments in epithelial cells).

- *Hemidesmosome*. Represent a "half desmosome"; these junctions anchor the basal surface of the cell to basal lamina.

- *Junctional complex*. Consists of the zonula occludens, zonula adherens, and desmosomes; because these structures cannot be resolved as separate structures at the light microscopic level, they appear as a single, bar-shaped, dark region at the apical corners of adjacent cells. The term *terminal bar* was used by early microscopists to define the zonula occludens and zonula adherens at the light microscopic level.
- *Gap junction*
  - Gap junctions consist of *connexons*, six transmembrane proteins clustered in a rosette that defines a central pore. Connexons from adjacent cells abut one another, forming a continuity between cells.
  - Provides metabolic and electrical continuity (coupling) via the pores between cells

Glandular Epithelial Tissues

General Considerations

- Develop from or within a lining or covering epithelium
- Secretory cells may
  - Differentiate but remain in the lining epithelium
  - Invaginate into the underlying connective tissue and remain attached to the lining epithelium
  - Invaginate into the underlying connective tissue but lose their connection to the epithelium

Exocrine vs. Endocrine Glands

- Major classification of glands, which is based on the method by which their secretory product is distributed
- *Exocrine glands*
  - Secretory products are released onto an external or internal epithelial surface, either directly or via a duct or duct system.
  - Secretory cells display polarized distribution of organelles.
- *Endocrine glands*
  - No ducts; secretory products are released directly into the extracellular fluid where they can affect adjacent cells (paracrine secretion) or enter the bloodstream to influence cells throughout the body (endocrine secretion).

- No polarization of organelles, except the thyroid gland and enteroendocrine cells of the digestive tract
- Secretory products are called hormones.

METHODS OF PRODUCT RELEASE FROM GLANDULAR CELLS

- *Merocrine*. Secretory product is released by exocytosis of contents contained within membrane-bound vesicles. This method of release is used by both exocrine and endocrine glands. Examples are digestive enzymes from pancreatic acinar cells and insulin from pancreatic islet cells.
- *Apocrine*. Secretory material is released in an intact vesicle along with some cytoplasm from the apical region of the cell. This method of release is used by exocrine glands only. An example is the lipid component of the secretory product of the mammary gland.
- *Holocrine*. Entire cell is released during the secretory process. Cells that are released may be viable (oocyte or sperm) or dead (sebaceous glands). This method of release is used by exocrine glands only.
- *Diffusion*. Secretory product passes through the cell membrane without the formation of secretory granules. Examples are steroid hormones. This method of release is used by endocrine glands only.

TYPES OF SECRETORY PRODUCTS

- Exocrine glands
  - *Mucus*. Thick, viscous, glycoprotein secretion
    - Secretory cells are usually organized into tubules with wide lumens.
    - Cytoplasm appears vacuolated, containing mucigen that, upon release, becomes hydrated to form mucus.
    - Nucleus is flattened and located in the base of the cell.
  - *Serous*. Thin, watery, protein secretion
    - Secretory cells are usually organized into a flask-shaped structure with a narrow lumen, called an acinus.
    - Cytoplasm contains secretory granules.
    - Nucleus is round and centrally located in the cell.
  - Special
    - *Lipid*. Oily secretion (sebum) from sebaceous glands and lipid portion of milk from the mammary gland.

- Sweat. Hypotonic, serous secretion that is low in protein content.
- Cerumen. A waxy material formed by the combination of the secretory products of sebaceous and ceruminous glands with desquamated epidermal cells in the auditory canal

➢ Endocrine glands

- *Protein* (e.g., insulin) or amino acid derivatives (e.g., thyroxine)
- *Steroid* (e.g., estrogen and testosterone)

Classification of Exocrine Glands

➢ *Unicellular glands*. Individual cells located within an epithelium, such as *goblet cells* that secrete mucus

➢ Multicellular glands

- *Sheet gland*. Composed of a surface epithelium in which every cell is a mucus secreting cell. A sheet gland is unique to the lining of the stomach.
- The remaining multicellular glands are classified according to:
  - The shape(s) of the secretory units
    - Presence of *tubules* only
    - Presence of *acini* (singular, *acinus*) or *alveoli* (singular, *alveolus*) (these two terms are synonymous), which are flask-shaped structures
    - Presence of both tubules and acini
  - The presence and configuration of the duct
    - *Simple*. No duct or a single, unbranched duct is present.
    - *Compound*. Branching duct system
  - Classification and types of multicellular glands
    - *Simple tubular*. No duct; secretory cells are arranged like a test tube that connects directly to the surface epithelium (e.g., intestinal glands).
    - *Simple, branched tubular*. No duct; tubular glands whose secretory units branch (e.g., fundic glands of stomach)
    - *Simple, coiled tubular*. Long unbranched duct; the secretory unit is a long coiled tube (e.g., sweat glands).
    - *Simple, branched acinar (alveolar)*. Secretory units are branched and open into a single duct (e.g., sebaceous glands).
    - *Compound tubular*. Branching ducts with tubular secretory units (e.g., Brunner's gland of the duodenum)

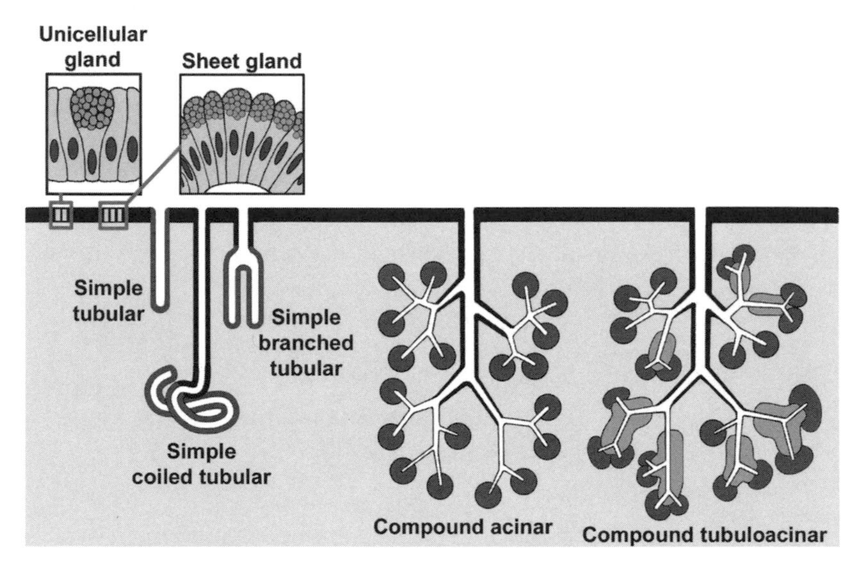

FIGURE 3.3. Types of glands based on their morphology.

- *Compound acinar (alveolar).* Branching ducts with acinar secretory units (e.g., parotid salivary gland)
- *Compound tubuloacinar (alveolar).* Branching ducts with both tubular and acinar secretory units (e.g., submaxillary salivary gland)

SPECIAL FEATURES OF SOME EXOCRINE GLANDS

- ➢ *Serous demilunes.* Consist of a "cap" of serous cells around the end of a mucous tubule; appear half-moon shaped in section
- ➢ *Myoepithelial cells.* Resemble smooth muscle cells in their fine structure but are of epithelial origin; prominent in sweat and mammary glands, they surround secretory units, lying inside the basement membrane, and aid in the expulsion of secretory products from the gland.

DUCT SYSTEM OF COMPOUND, EXOCRINE GLANDS

- ➢ *Intralobular ducts.* Contained within a lobule; simple cuboidal to columnar epithelium
- ➢ *Interlobular ducts.* Receive numerous intralobular ducts; located in the connective tissue between lobules; stratified columnar epithelium
- ➢ *Excretory (main) duct.* Macroscopic duct draining the entire gland

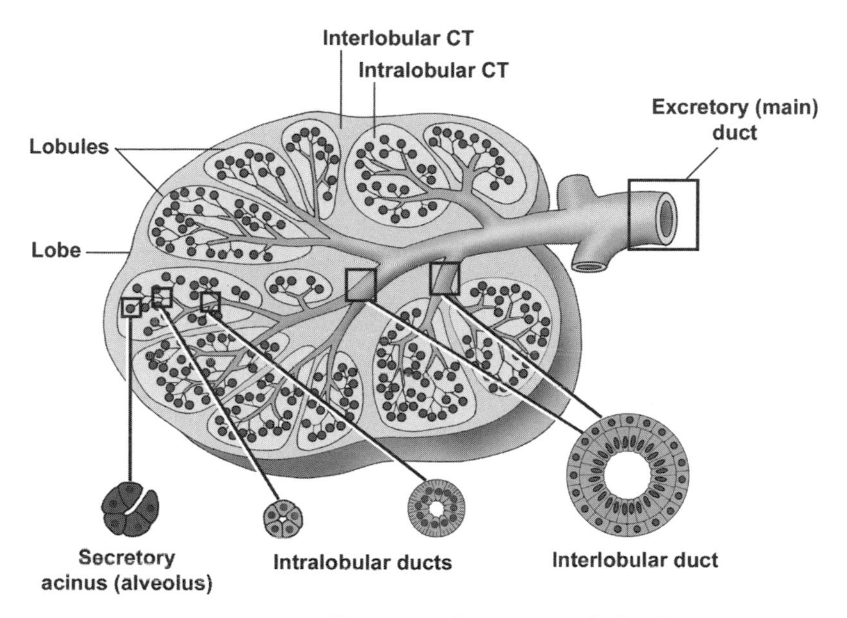

FIGURE 3.4. Structure of a compound gland.

Endocrine Glands (See also Chapter 16)

- No ducts; generally cells are not polarized
- Occurrence
  - Unicellular (e.g., enteroendocrine cells of the digestive tract); these cells do show polarity because they are located within an epithelium and secrete away from the free surface of the epithelium.
  - Small clusters of cells (e.g., islet of Langerhans in pancreas)
  - Organs (e.g., thyroid gland, adrenal gland)
- Secretory cells of multicellular glands are usually arranged as plates or cords. The thyroid gland, where the cells form fluid-filled spheres, is an exception to this pattern.
- Highly vascular with fenestrated capillaries
- Secretory products are called hormones. Hormones can be:
  - Derived from amino acids (e.g, thyroxine and epinephrine)
  - Peptides and proteins (e.g., insulin and oxytocin)
  - Steroids (e.g., testosterone and cortisol); steroid-secreting cells display mitochondria with tubular cristae and contain large amounts of lipid droplets and smooth endoplasmic reticulum.
- Secrete by the merocrine or diffusion methods only

Structures Identified in This Section

Lining and Covering Epithelial Tissues

- Types of epithelial membranes
  - Simple squamous epithelium
  - Simple cuboidal epithelium
  - Simple columnar epithelium
  - Pseudostratified epithelium
    - Basal cells
  - Stratified squamous, nonkeratinized epithelium
    - Basal cells
    - Squamous cells
  - Stratified squamous, keratinized epithelium
    - Basal cells
    - Keratin
    - Surface dead cells
  - Stratified columnar epithelium
  - Transitional epithelium (relaxed and stretched)
    - Dome cells
- Surface specializations
  - Actin filament bundles
  - Basal bodies
  - Brush border
  - Cilia
  - Microvilli
  - Stereocilia
  - Terminal web
- Basement membrane
  - Basal lamina
    - Lamina densa
    - Lamina lucida
  - Collagen fibrils
  - Reticular lamina
- Cell junctions
  - Desmosomes
  - Hemidesmosomes
  - Junctional complex
  - Terminal bars
  - Zonula adherens
  - Zonula occludens

Glandular Epithelial Tissues

- Goblet cell
  - Mucigen
- Sheet gland
- Simple tubular gland
- Simple branched tubular gland
  - Body of gland
  - Gastric pit
- Simple coiled tubular gland
  - Duct
  - Myoepithelial cell
  - Secretory portion
- Simple branched alveolar gland
  - Acinus (alveolus)
  - Duct
  - Lipid
- Compound acinar gland
  - Blood vessel
  - Interlobular connective tissue
  - Interlobular duct
  - Intralobular connective tissue
  - Intralobular duct
  - Lobule
  - Secretory granule
  - Serous acinus
    - Basal rough endoplasmic reticulum
    - Secretory granules
- Compound tubulo-alveolar gland
  - Interlobular connective tissue
  - Interlobular duct
  - Intralobular duct
  - Lobule
  - Mucous tubule
  - Secretory granule
  - Serous acinus (alveolus)
  - Serous demilune

CHAPTER

4

Connective Tissues

General Concepts

➢ Composition

- *Cells.* Each type of connective tissue has it own characteristic complement of one or more of a wide variety of cells.
- *Extracellular matrix.* Synthesized and secreted by resident "blast" cells specific for each connective tissue type (e.g., fibroblasts and chondroblasts); the matrix is composed of:
  - ◆ *Fibers.* Collagen, elastic and reticular
  - ◆ *Ground substance.* An amorphous substance that can exist as a liquid, gel, or flexible or rigid solid, conferring unique structural properties to each connective tissue.

➢ Functions

- Provides substance and form to the body and organs
- Defends against infection
- Aids in injury repair
- Stores lipids
- Provides a medium for diffusion of nutrients and wastes
- Attaches muscle to bone and bone to bone

Digital Histology: An Interactive CD Atlas with Review Text, by Alice S. Pakurar and John W. Bigbee
ISBN 0-471-64982-1

- Types of connective tissue. Classified by the relative abundance, variety, and content of their components
  - *Connective tissue proper*
  - *Cartilage*
  - *Bone*
  - *Special*. Includes adipose, elastic, reticular, and mucoid connective tissues as well as blood and hematopoietic tissue

THE CONNECTIVE TISSUES: CONNECTIVE TISSUE PROPER

GENERAL CONCEPTS

- Connective tissue proper comprises a very diverse group of tissues, both functionally and structurally.
  - Structural functions of connective tissue proper
    - Forms a portion of the wall of hollow organs and vessels and the stroma of solid organs
    - Forms the stroma of organs and subdivides organs into functional compartments
    - Provides padding between and around organs and other tissues
    - Provides anchorage and attachment (e.g., muscle insertions)
  - Provides a medium for nutrient and waste exchange
  - Lipid storage in adipocytes
  - Defense and immune surveillance function via lymphoid and phagocytic cells
- All connective tissues are composed of two basic components, which vary widely among different types of connective tissue. The components of connective tissues are:
  - Cells (e.g., fibroblasts and macrophages)
  - Extracellular matrix
    - Fibers (e.g., collagen and elastic fibers)
    - Ground substance

CELLS OF CONNECTIVE TISSUE

- Connective tissue cells can be subdivided into two major groups. *Resident cells* are derived from mesenchyme and are continuously present in the tissue (e.g., fibroblasts and adipocytes). *Migratory cells* enter and leave the blood stream to migrate through and function in connective tissues (e.g., neutrophils and macrophages [monocytes]).

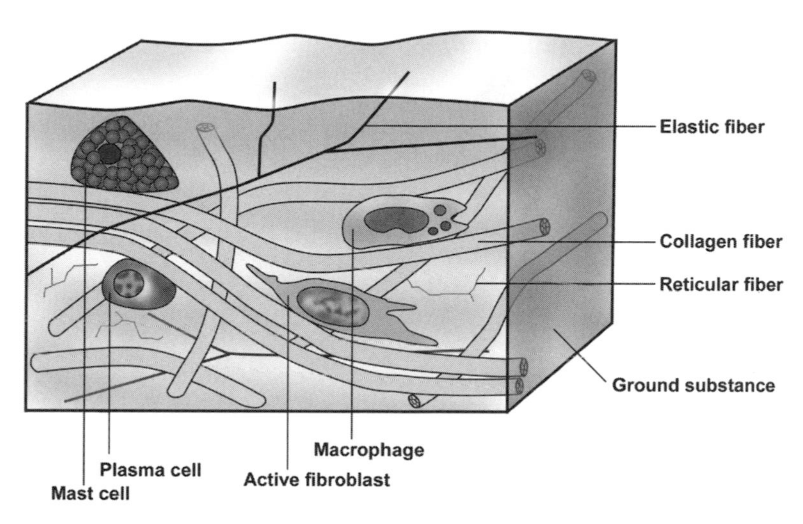

FIGURE 4.1. Components of connective tissue proper.

- Fibroblasts
  - Synthesize and maintain fibers and ground substance
  - Major resident cell in connective tissue proper
  - Active and inactive fibroblasts
    - *Active fibroblast*
      - Large, euchromatic, oval nucleus
      - Cytoplasm not usually visible but contains abundant rough endoplasmic reticulum and Golgi
      - Elongated, spindle-shaped cells
      - High synthetic activity
    - *Inactive fibroblast*
      - Small, heterochromatic, flattened nucleus
      - Reduced cytoplasm and organelles
      - Low synthetic activity
- *Macrophages*
  - Derived from blood monocytes; monocytes enter connective tissue from the bloodstream and rapidly transform into macrophages that function in phagocytosis, antigen processing, and cytokine secretion.
  - Comprise the mononuclear phagocyte system of the body; include Kupffer cells in the liver, alveolar macrophages in the lung,

microglia the central nervous system, Langerhan's cells in the skin, and osteoclasts in bone marrow

- Structure
  - Heterochromatic, oval nucleus with an indentation in the nuclear envelope and marginated chromatin
  - Cytoplasm usually not visible unless it contains phagocytosed material

➢ *Mast cells*

- Mediate immediate hypersensitivity reaction and anaphylaxis by releasing immune modulators from cytoplasmic granules, in response to antigen binding with cell surface antibodies
- Structure
  - Round to oval-shaped cells
  - Round, usually centrally located nucleus
  - Well-defined cytoplasm filled with secretory granules containing immune-modulatory compounds (e.g., histamine and heparin)

➢ *Plasma cells*

- Secrete antibodies to provide humoral immunity
- Derived from B-lymphocytes
- Structure
  - Oval-shaped cells
  - Round, eccentrically located nucleus with heterochromatin clumps frequently arranged like the numerals on a clock-face
  - Basophilic cytoplasm due to large amounts of rough endoplasmic reticulum
  - Well-developed Golgi complex appears as a distinct, unstained region in the cytoplasm near the nucleus and, for that reason, is often referred to as a "negative Golgi."

➢ *Adipose cells (adipocytes, fat cells)*

- Store lipids
- Types
  - *Yellow fat (unilocular)*
    - Each cell contains a single droplet of neutral fat (triglycerides) for energy storage and insulation.
    - Minimal cytoplasm, present as a rim around the lipid droplet
    - Flattened, heterochromatic, crescent-shaped nucleus that conforms to the contour of the lipid droplet

      - Can occur singly, in small clusters or forming a large mass, which is then referred to as adipose connective tissue
    - *Brown fat (multilocular)*
      - Cells contain numerous, small lipid droplets.
      - Large numbers of mitochondria
      - Present mostly during early postnatal life in humans, abundant in hibernating animals for heat production

➢ *White blood cells (WBCs, leukocytes)*

- These cells enter and leave the blood stream to migrate through, and function in, connective tissues. The most common WBCs encountered in connective tissue proper are lymphocytes, neutrophils, and eosinophils. For a complete discussion of blood cells see "Blood and Hematopoiesis" (Chapter 6).
  - *Lymphocytes (T and B lymphocytes)*
    - Small spherical cells with sparse cytoplasm and a round heterochromatic nucleus, often with a small indentation
    - B cells enter connective tissue where they transform into plasma cells and secrete antibodies. T cells are primarily located in lymphatic tissues and organs; however, T cells can be present in connective tissue proper under certain circumstances (e.g., organ transplantation).
  - *Neutrophils (polymorphonuclear leukocytes, PMNs)*
    - Spherical cells with a heterochromatic nucleus with three to five lobes
    - Pale-staining cytoplasmic granules
    - Highly phagocytic cells that are attracted to sites of infection
  - *Eosinophils*
    - Spherical cells with a bilobed nucleus
    - Cytoplasmic granules stain intensely with eosin.
    - Modulate the inflammatory process

EXTRACELLULAR MATRIX

➢ *Fibers*

Fiber type	Composition	Properties
Collagen	Collagen I, II	Inelastic, eosinophilic
Reticular	Collagen III	Inelastic, branched, argyrophilic
Elastic	Elastin	Elastic, eosinophilic

- *Collagen fibers*
  - *Tropocollagen*
    - Basic collagen molecule consisting of three alpha subunits intertwined in a triple helix; collagen types are distinguished by their subunit composition.
    - Produced by fibroblasts and other matrix-forming cells
    - Secreted into the matrix, where they spontaneously orient themselves into fibrils with a 64-nm repeating banding pattern
  - Major collagen types
    - *Type I.* Fibrils aggregate into fibers and fiber bundles; most widespread distribution; "interstitial collagen."
    - *Type II.* Fibrils do not form fibers; present in hyaline and elastic cartilages
    - *Type III.* Fibrils aggregate into fibers; present surrounding smooth muscle cells and nerve fibers; forms the stroma of lymphatic tissues and organs
    - *Type IV.* Chemically unique form of collagen that does not form fibrils; major component of the basal lamina
- *Elastic fibers*
  - Composed primarily of elastin; produced by fibroblasts
  - Elastin forms the central amorphous core of the fiber, which is surrounded by microfibrils.
  - Unique chemical properties of elastin provide for elasticity.
  - Elastic fibers occur in nearly all connective tissues in varying amounts and are intermixed with collagen fibers. When present exclusively, they constitute elastic connective tissue.
  - Frequently difficult to differentiate from collagen with conventional stains
- *Reticular fibers*
  - Collagen type III fibers
  - Highly glycosylated and stain with silver (argyrophilic)
  - When they are the major fiber fiber type (e.g., in the stroma of lymphoid organs), they constitute reticular connective tissue.

➢ *Ground substance*

- Functions
  - Forms a gel-like matrix of variable consistency in which cells and fibers are embedded

    - Provides a medium for passage of molecules and cells migrating through the tissue
    - Contains adhesive proteins that regulate cell movements
  - Components
    - *Tissue fluid*. Contains salts, ions and soluble protein
    - *Glycosaminoglycans (GAGs)*
      - Long, unbranched polysaccharides composed of repeating disaccharide units, which are usually sulfated
      - Large negative charge of the sugars attracts cations, resulting in a high degree of hydration. The matrix formed ranges from a liquid passageway to a viscous shock absorber.
      - GAGs are generally attached to proteins to form *proteoglycans*.
      - *Proteoglycan aggregate*. Many proteoglycans are attached to hyaluronic acid, which is itself a glycosaminoglycan.
    - *Adhesive glycoproteins*. For example, fibronectin and laminin.

CLASSIFICATION OF CONNECTIVE TISSUE

CONNECTIVE TISSUE PROPER

- Loose (areolar)
  - Highly cellular, numerous cell types present
  - Fewer and smaller caliber collagen fibers compared with dense
  - Abundant ground substance, allows for diffusion of nutrients and wastes
  - Highly vascularized
  - Provides padding between and around organs and tissues
- Dense
  - Fewer cells, mostly fibroblasts
  - Highly fibrous with larger caliber collagen fibers, provides strength
  - Minimal ground substance
  - Poorly vascularized
  - Types
    - *Dense, irregular connective tissue*. Fiber bundles arranged in an interlacing pattern; forms the capsule of organs and the dermis of the skin
    - *Dense regular connective tissue*. Parallel arrangement of fiber bundles; restricted to tendons and ligaments

CONNECTIVE TISSUES WITH SPECIAL PROPERTIES

- *Adipose connective tissue*. Consists of accumulations of adipocytes that are partitioned into lobules by septa of connective tissue proper. Provides energy storage and insulation
- *Blood and hematopoietic (blood-forming) tissues* (Chapter 6)
- *Elastic connective tissue*. Regularly arranged elastic fibers or sheets (e.g., the vocal ligament)
- *Reticular connective tissue*. A loosely arranged connective tissue whose fibers are reticular fibers. Forms the stroma of hematopoietic tissue (e.g., bone marrow) and lymphoid organs (e.g., lymph node and spleen).
- *Mucoid connective tissue*. Embryonic connective tissue with abundant ground substance and delicate collagen fibers; present in the umbilical cord

SUPPORTIVE CONNECTIVE TISSUES

- Cartilage (Chapter 5)
- Bone (Chapter 5)

STRUCTURES IDENTIFIED IN THIS SECTION

Connective tissue types
- Adipose connective tissue
- Dense, irregular connective tissue
- Dense, regular connective tissue
- Loose or areolar connective tissue
- Reticular connective tissue

Connective tissue fiber types
- Collagen fibers/bundles
- Collagen fibrils
- Elastic fibers
- Reticular fibers

Cell types
- Active fibroblast with a euchromatic nucleus
- Adipocyte
- Eosinophil
- Inactive fibroblast with a heterochromatic nucleus
- Lymphocyte
- Macrophage
- Mast cell
- Negative Golgi
- Neutrophil
- Plasma cell

CHAPTER

5

SUPPORTING CONNECTIVE TISSUES

CARTILAGE

OVERVIEW

- ➢ Composition is similar to that of all connective tissues
  - *Cells*
  - *Extracellular matrix* consisting of fibers, ground substance, and tissue fluid
- ➢ Cartilage is avascular and possesses no lymph vessels or nerves.
- ➢ Types of cartilage
  - Hyaline. Provides nonrigid support
  - Elastic. Provides support with large amount of flexibility
  - Fibrocartilage. Provides strength under stress

COMPONENTS OF CARTILAGE

- ➢ *Perichondrium*. Connective tissue surrounding cartilage tissue. Layers include:
  - *Fibrous layer*. Outer portion, composed of dense connective tissue, serves as a source of reserve cells for the chondrogenic layer

Digital Histology: An Interactive CD Atlas with Review Text, by Alice S. Pakurar and John W. Bigbee
ISBN 0-471-64982-1

  - *Chondrogenic layer*. Inner, more cellular portion contains chondroblasts and blends imperceptibly into cartilage tissue proper.
- Cells
  - *Chondroblasts*
    - Lie on the surface of cartilage in the chondrogenic layer of perichondrium
    - Secrete extracellular matrix around themselves, thus becoming chondrocytes
  - *Chondrocytes*
    - Are chondroblasts that have surrounded themselves with matrix
    - Lie within cartilage in potential spaces called lacunae
    - Secrete and maintain extracellular matrix
    - Are frequently located in *isogenous groups*, a cluster of chondrocytes, resulting from the proliferation of a single chondrocyte
- *Extracellular matrix*. Both flexible and noncompressible
  - Composition
    - *Fibers*. Collagen fibrils and fibers predominate in hyaline cartilage and fibrocartilage, respectively; elastic fibers predominate in elastic cartilage.
    - *Ground substance*. Tissue fluid surrounds proteoglycan aggregates bound to collagen fibers. Collectively, these form a firm gel, which resists compressive forces.
  - Subdivisions
    - *Territorial matrix* immediately surrounds chondrocytes. This matrix stains more intensely with hematoxylin due to the high concentration of proteoglycans.
    - *Interterritorial matrix* is the lighter-staining matrix outside the territorial matrix and between isogenous groups.

GROWTH OF CARTILAGE

- *Appositional* growth occurs at the surface of cartilage. New cartilage is added (apposed) to the surface of preexisting cartilage by the activity of chondroblasts lying in the chondrogenic layer of the perichondrium.
- *Interstitial* growth occurs from within cartilage tissue. Chondrocytes produce additional matrix and divide, forming isogenous groups.

TYPES OF CARTILAGE

- *Hyaline cartilage*
  - Is the most common cartilage type and is hyaline (glassy) in appearance
  - Contains collagen type II fibers that have the same refractive index as ground substance and, therefore, are not visible with the light microscope by conventional staining methods
  - Stains blue with conventional dyes, due to the relative abundance of its ground substance matrix
  - Possesses numerous isogenous groups
  - Function and distribution. Forms most of the cartilages of the body, comprises the fetal skeleton, attaches ribs to the sternum, forms epiphyseal plates, and lines articular surfaces. (The lack of a perichondrium on the articular cartilages provides a smooth, glassy articular surface.)
- *Elastic cartilage*
  - Has a visible network of interlacing elastic fibers in addition to collagen type II fibers
  - Possesses fewer isogenous groups than does hyaline cartilage
  - Function and distribution. Pinna of ear, epiglottis, smaller laryngeal cartilages (i.e., present where flexibility and support are necessary)
- *Fibrocartilage*
  - Is a functional and structural intermediate between hyaline cartilage and the dense connective tissues
  - Possesses abundant collagen type I fibers, arranged in either a regular or irregular configuration. These collagen fibers cause this cartilage to stain pink with eosin.
  - Has minimal ground substance. The ground substance that is present is usually located immediately around the chondrocytes.
  - Possesses few isogenous groups
  - Combines great tensile strength with flexibility
    - Frequently found where a tendon or a ligament attaches to a bone (regular arrangement of fibers)
    - Located in the pubic symphysis and knee cartilages (irregular fiber arrangement)

Regressive Changes in Cartilage

- Occur in cartilage more frequently than in many other tissues
- Regressive changes also occur in the hyaline cartilage of the epiphyseal plate and represent critical steps in endochondral bone formation.
- Stages of regression
  - Chondrocytes hypertrophy and secrete alkaline phosphatase that provides a calcifiable matrix.
  - Calcium phosphate is deposited in the matrix, prohibiting diffusion of nutrients to the chondrocytes.
  - Chondrocytes die, leaving behind empty lacunae and the calcified matrix.

Structures Identified in This Section

Calcified cartilage matrix
Chondroblasts
Chondrocytes
Collagen bundles
Elastic cartilage
Elastic fibers
Fibrocartilage
Ground substance
Hyaline cartilage
Hypertrophy
Interterritorial matrix
Isogenous group
Lacunae
Perichondrium
Perichondrium, chondrogenic layer
Perichondrium, fibrous layer
Territorial matrix

Supporting Connective Tissues: Bone

General Concepts

- Bone
  - Provides structural support, giving shape and form to the body
  - Provides movement through the insertion of muscles
  - Serves as a stored source for calcium and phosphate
  - Contains bone marrow (myeloid tissue)
- Histological preparation of bone
  - Ground bone preparation. Unpreserved bone is ground to a thinness where light can be transmitted through it. Because no preservation has occurred, neither cells nor organic matrix survive.

Lamellae, lacunae, canaliculi, and general architecture of inorganic matrix are well displayed.

- Decalcified bone. Cells are fixed (preserved) and inorganic matrix removed by decalcification. Good detail of organic matrix (cells, periosteum, etc.) is maintained, but lamellae and inorganic matrix are difficult to distinguish.

Gross Appearance of Bone, Macroscopic Structure

➢ *Compact bone.* Appears as a solid mass to the naked eye, covering the exterior of bones and forming the shaft of long bones.

➢ *Spongy* or *cancellous bone.* Gross appearance is like a sponge, with a labyrinth of bony spicules and intervening spaces that are filled with loose connective tissue or red marrow and at least one blood vessel. Spongy bone is located in the interior of bones.

Architecture of a Long Bone

➢ The *diaphysis* (shaft), composed of compact bone, is hollow and is usually lined by a thin band of spongy bone.

➢ An *epiphysis*, the knob at either end of the diaphysis, is composed of a thin rim of compact bone. The spongy bone in its interior houses red marrow.

➢ *Metaphysis.* Flared region between diaphysis and epiphysis.

➢ *Epiphyseal plate.* Hyaline cartilage separating epiphysis and metaphysis in growing bones. Growth in bone length occurs as hyaline cartilage in the epiphyseal plate goes through various stages of regression, providing a framework on which bone is deposited. When the hyaline cartilage in the epiphyseal plate is exhausted, growth stops. The epiphysis and metaphysis fuse in the adult, leaving an epiphyseal line as a remnant of the epiphyseal plate.

➢ Marrow

- *Red marrow*, found in all bones of the fetus, is restricted to spongy bone areas of selected bones in the adult and contains hematopoietic tissue that forms blood cells.
- *Yellow marrow*, found in the shafts of long bones in the adult, consists mainly of adipose connective tissue that retains the potential to become red marrow under hemorrhagic stress.

➢ *Articular cartilage* is composed of hyaline cartilage and covers articular surfaces of bone. This cartilage does not possess a perichondrium; the glassy, smooth cartilage provides a good articulating surface.

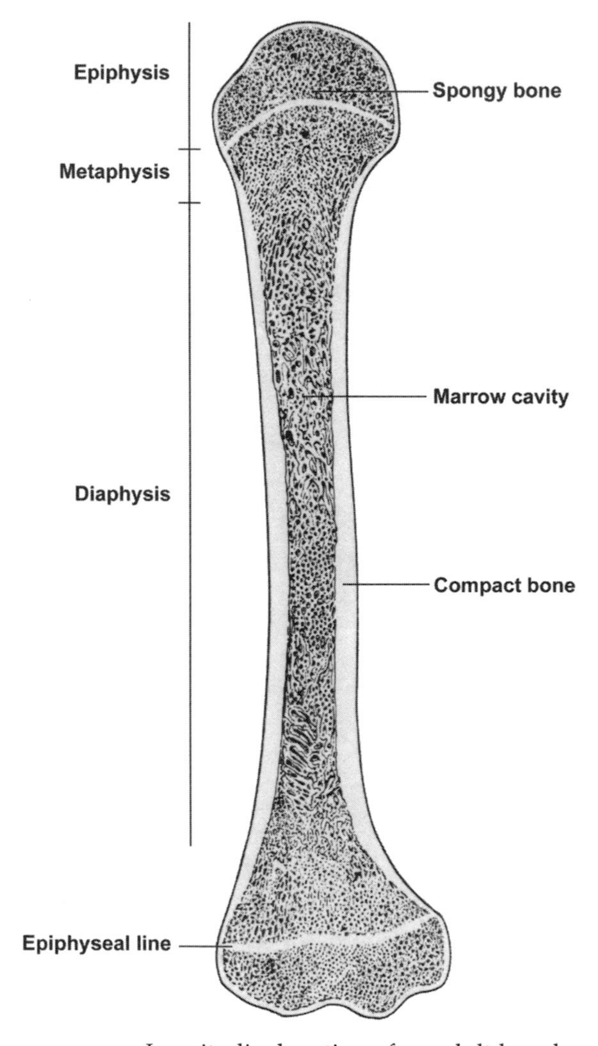

FIGURE 5.1. Longitudinal section of an adult long bone.

COMPONENTS OF BONE

➢ *Extracellular matrix*

- *Organic portion, osteoid*. Secreted by osteoblasts
  - Collagen type I *fibers* comprise the majority of the organic matrix. Their predominance causes bone to stain pink with eosin.
  - *Ground substance* is minimal, composed of glycosaminoglycans such as chondroitin sulfate, keratan sulfate, and some glycoproteins that avidly bind calcium.

- *Inorganic portion.* Calcium and phosphate, in the form of hydroxyapatite crystals, are deposited along the collagen fibrils and form 50% of the dry weight of bone. This ossified matrix renders bone impermeable to diffusion of nutrients and requires that bone be well vascularized.

➢ Cells

- *Osteoblasts*
  - Located on all exterior surfaces of bone as the innermost portion of the periosteum or in the endosteum lining all interior bony surfaces
    - Inactive osteoblasts are flattened cells with heterochromatic nuclei.
    - Active osteoblasts are stellate and contain organelles necessary for protein, primarily collagen, production. These cells synthesize high levels of alkaline phosphatase.
  - Function to synthesize bone
    - Secrete osteoid first
    - In the presence of alkaline phosphatase, osteoblasts facilitate the deposition of calcium phosphate, thus mineralizing the osteoid.
- *Osteocytes*
  - Are osteoblasts that have completely surrounded themselves by bony matrix and, therefore, must lie within, rather than on, bone tissue. These flattened, inactive cells lie in *lacunae* (spaces) in the bone and extend long processes from the cell body. These processes lie in narrow tunnels called *canaliculi* and connect, via gap junctions, with adjacent osteocytes and/or osteoblasts at the bone surface.
  - Function to transport materials between blood and bone and to maintain surrounding matrix; they do not divide or secrete matrix.
- *Osteoclasts*
  - Are large cells with 15–20 or more nuclei and vacuolated, frothy cytoplasm. A ruffled border, the highly enfolded cell membrane facing the bone, is the site of bone resorption.
  - Are located on internal surfaces as part of the endosteum or on external surfaces as part of the osteogenic layer of the periosteum. Osteoclasts lie in depressions in the bone, *Howship's lacunae*, which form as osteoclasts resorb bone.
  - Resorb bone via the acid phosphatase and proteolytic enzymes they secrete

- Surface coverings
  - *Periosteum*. Double layer of connective tissue surrounding the outer surface of bones, except for articular surfaces
    - Layers
      - *Fibrous layer*. Outer layer of dense connective tissue that serves as a reserve-cell source for the osteogenic layer
      - *Osteogenic layer*. Inner, more cellular layer, contains osteoblasts and osteoclasts. Site of bone deposition and resorption, respectively.
    - Well vascularized and richly innervated
  - *Endosteum*
    - Is composed of a single row of osteoblasts, osteoclasts, and/or osteo-progenitor cells that lines all interior surfaces of bone except for lacunae and canaliculi
    - Serves as a means of bone growth and/or resorption

MICROSCOPIC APPEARANCE OF BONE (RELATED TO THE AGE OF A BONE)

- *Woven* or *immature bone* is the first bone deposited.
  - May be either spongy or compact
  - Referred to as woven bone because fibers are deposited in a random array
  - Contains osteocytes that are more numerous and spherical than those of lamellar bone. These osteocytes are not in any orderly arrangement.
  - Is less well mineralized than lamellar bone and, therefore, appears bluer than lamellar bone with hematoxylin and eosin stains
  - Is usually resorbed and replaced by lamellar bone
- *Lamellar* or *mature bone*
  - Replaces most woven bone or may be deposited *de novo*
  - May be either spongy or compact
  - Is referred to as lamellar bone because the matrix is deposited in layers or lamellae
  - Fibers are deposited in parallel array within a lamella.
  - Osteocytes are fewer and flatter than those in woven bone and are organized in rows between the lamellae.
  - Better mineralized than woven bone

- Bone is not a static structure and is constantly being resorbed and reconstructed. Therefore, lamellar bone is also resorbed and reconstructed throughout life.

ARCHITECTURE OF ADULT, COMPACT LAMELLAR BONE

➢ *Outer circumferential lamellae.* Stacks of lamellae extend at least partially around the outer circumference of a long bone. Deposition of these lamellae by the periosteum results in increased thickness in the wall of the diaphysis.

➢ *Inner circumferential lamellae.* Stacks of lamellae extend at least partially around the inner circumference of a long bone facing the marrow cavity. Deposition of these lamellae by the endosteum results in increased thickness of the wall of the diaphysis.

➢ *Haversian systems, osteons*

- Primary structures of compact lamellar bone
- Cylinders of concentric lamellae, deposited by endosteum, that run parallel to the long axis of a bone
- Composition
  - ◆ Central *Haversian canal*
    - ■ Appears round in cross-section with a smooth periphery
    - ■ Contains a blood vessel(s) and loose connective tissue
    - ■ Is lined with an endosteum
  - ◆ Concentric lamellae (4–20) surround the Haversian canal.
    - ■ Collagen fibers are in parallel alignment within a single lamella, wrapping helically around the Haversian canal.
    - ■ Pitch of the helix varies with each lamella in the osteon.
- Provides great strength to a long bone
- An osteon is formed by the centripetal deposition of the concentric lamella (i.e., outer lamella is the oldest).

➢ Additional lamellae/structures associated with adult, compact lamellar bone

- *Interstitial lamellae.* Portions of Haversian systems that remain after resorption of the rest of the osteon. These lamellae are interposed between other, complete Haversian systems.
- *Volkmann's canals.* Channels oriented perpendicularly between adjacent Haversian canals, interconnecting these canals with each other and with the surfaces of bone. Volkmann's canals contain blood vessels that transport blood from the surface of bone to blood vessels within Haversian canals.

- *Cement lines*. Thin, refractive lines that are collagen poor and stain, therefore, with hematoxylin. Cement lines are located:
  - Around Haversian systems, demarcating where resorption stopped and the formation of a new osteon began
  - Beneath and between circumferential lamellae, denoting where deposition of lamellae halted for a period of time and then began again

BONE GROWTH, DEPOSITION, AND RESORPTION

BONE GROWTH

- New, adult bone is always laid down on preexisting bone or cartilage.
- Bone growth is always appositional, with either endosteum or periosteum laying down lamellae of bone. Interstitial growth is impossible in bone because its rigid, ossified matrix does not allow osteocytes to secrete additional matrix or to divide.

BONE DEPOSITION

- Newly deposited bone assumes the shape of the bone or cartilage on which it is deposited
- In spongy, lamellar bone, new lamellae are laid down by osteoblasts in the endosteum located at the periphery of each trabecula, thus increasing its thickness.
- In compact lamellar bone, new lamellae are laid down either as outer circumferential lamellae by osteoblasts in the periosteum or as inner circumferential lamellae and Haversian systems (osteons) by the endosteum.

BONE RESORPTION

- Removal of bone by osteoclasts for remodeling during growth and/or to mobilize calcium throughout life
- Resorption process
  - Osteoclasts on the periosteal and endosteal surfaces resorb bone from bone surfaces.
  - *Resorption canal*
    - Is a cylindrical, longitudinal tunnel formed as compact bone on the interior of bone is resorbed

- ◆ Appears in cross-section as an irregularly shaped, bony surface lined with an endosteum containing osteoclasts
- ◆ Usually extends past cement lines, eroding through portions of several osteons. Therefore, remnants of resorbed osteons may surround the resorption canal.
- ◆ Is not lined by concentric lamellae as are osteons
- ◆ When resorption stops, osteoblasts begin filling in a resorption canal by centripetal (from outside to inside) deposition of new lamellae, forming a new osteon. The newest lamella of this secondary osteon is the one adjacent to the Haversian canal.
- ◆ Remains of partially resorbed Haversian systems around this secondary osteon are called interstitial lamellae.

Bone Formation (Ossification)

Intramembranous Bone Formation

- ➢ Definition. Bone formation by a connective tissue membrane. No cartilage precedes this bone formation. Bone formed may be woven or lamellar, spongy or compact.
- ➢ Connective tissue membranes involved in intramembranous ossification include mesenchyme in the fetus and periosteum or endosteum in both the fetus and the adult.
- ➢ Occurrence of intramembranous bone
  - ● Bone laid down by mesenchyme forming the flat bones of the skull and part of the mandible
  - ● Bone laid down by the periosteum or endosteum
- ➢ Types of intramembranous ossification
  - ● Ossification from mesenchyme in the fetus
    - ◆ Mechanism of ossification
      - ■ Mesenchymal cells cluster and differentiate into osteoblasts that secrete organic matrix (osteoid) around themselves. This matrix becomes mineralized, thereby forming bone.
      - ■ Bone formed is woven, spongy bone.
    - ◆ Many areas of this spongy, woven bone are converted to compact, lamellar bone by the filling in of the spaces between trabeculae with osteons.
    - ◆ Other areas of this spongy bone are not converted to compact, however, such as the spongy bone forming the diploe of flat bones of the skull.

- Ossification from a connective tissue membrane, such as periosteum or endosteum
  - Mechanism of ossification. Osteoblasts in the endosteum or in the osteogenic layer of the periosteum secrete and lay down lamellae of bone.
  - Lamellae conform to the shape of the bone or cartilage on which they are deposited:
    - Lamellae deposited around a cylindrical cavity form an osteon.
    - Circumferential lamellae form on the inner and outer surfaces of bone from the endosteum or periosteum, respectively.
    - Endosteum adds lamellae to trabeculae of spongy bone.

ENDOCHONDRAL BONE FORMATION

- Formation of bone by replacement of a preexisting hyaline cartilage template. The cartilage must first undergo regressive changes that produce a framework upon which bone is deposited (ossification).
- Bones formed endochondrally include bones at the base of the skull, long bones, vertebrae, pelvis, ribs.
- Events occurring before ossification begins
  - Hyaline cartilage template of the future bone is formed in the fetus. This cartilage is surrounded by a perichondrium and enlarges by appositional and interstitial growth as the fetus grows.
  - Regressive changes begin in cartilage cells in the central, diaphyseal region of the template at what will become the primary center of ossification.
    - Chondrocytes mature, greatly hypertrophy at the expense of surrounding matrix, and begin to secrete alkaline phosphatase.
    - The presence of alkaline phosphatase leads to the calcification of the cartilage matrix, making it impermeable to metabolites.
    - Chondrocytes die, leaving behind their lacunae separated by spicules of calcified cartilage matrix.
    - The oxygen supply to the fetus is increasing as the fetal circulatory system becomes functional, supplying blood to the hyaline cartilage template of the future bone.
- Stages of ossification
  - Formation of a *periosteal band* or *collar*
    - Around the middle of the shaft of the cartilage template, the chondroblasts differentiate into osteoblasts and begin secreting

a bony, rather than a cartilaginous, band called the periosteal band or collar. This cylinder of bone is formed by intramembranous ossification because it does not replace cartilage that has gone through regressive changes. The perichondrium surrounding the periosteal collar is now called a periosteum.

- ◆ The remainder of the cartilage template is surrounded by a perichondrium.

- *Primary center of ossification*
  - ◆ One of the fetal arteries, called the *periosteal bud,* and its surrounding mesenchymal cells, penetrate the diaphyseal region of the cartilage template into the area of the degenerating calcified cartilage.
  - ◆ Mesenchymal cells accompanying the artery differentiate into osteoblasts that deposit bone on the spicules of the calcified cartilage framework. Resulting spicules consist of:
    - ■ A core of calcified cartilage that stains blue with hematoxylin
    - ■ An outer perimeter of woven bone that stains pink with eosin
  - ◆ Some of the spicules of cartilage and bone are resorbed to form the future marrow cavity.
  - ◆ This cartilage degeneration–bone deposition process continues toward either epiphysis, becoming more organized into discrete zones, and forming the epiphyseal plate.
    - ■ *Resting zone* of "normal" hyaline cartilage
    - ■ *Zone of proliferation* where isogenous groups of chondrocytes actively divide, forming linear isogenous groups. This zone maintains cartilage thickness.
    - ■ *Zone of maturation* and *hypertrophy* of chondrocytes, with the production of alkaline phosphatase, and the subsequent *calcification* of the cartilage matrix
    - ■ *Zone of degeneration* where chondrocytes die, leaving empty lacunae surrounded by vertically oriented spicules of calcified cartilage
    - ■ *Zone of ossification* where bone is deposited on calcified cartilage spicules
    - ■ *Zone of resorption* where calcified cartilage–bone spicules are resorbed to form the marrow space

- *Secondary center of ossification* occurs in each epiphysis; ossification follows a similar pattern as that at the primary center except:
  - ◆ No periosteal band is formed.

    - ◆ Ossification occurs in a radial manner from the original center of the secondary center of ossification.
    - ◆ Bone resorption does not occur; thus, spongy bone permanently fills the epiphyses.
    - ◆ Ossification does not replace articular cartilage.
- ➢ Growth in length continues from epiphyseal plates, which:
  - Are established by formation of the primary and secondary centers of ossification
  - Are composed of hyaline cartilage showing the zonations described above
  - Are located between each epiphysis and metaphysis
  - Maintain a constant thickness throughout growth due to equivalent activity in the zones of proliferation and resorption
  - Are depleted at appropriate developmental stages as cartilage proliferation stops and the epiphyseal plate can no longer perpetuate itself. Spongy bone replaces the epiphyseal plate, leaving an epiphyseal line as its remnant. This process is referred to as closure of the epiphyseal plate.

STRUCTURES IDENTIFIED IN THIS SECTION

Bone Tissue
- Blood vessels
- Bone marrow
- Canaliculi
- Compact bone
- Decalcified bone
- Ground bone
- Howship's lacunae
- Intercellular matrix
- Lacunae
- Lamellae of bone
- Lamellar bone
- Loose connective tissue
- Organic matrix
- Osteoblasts, active
- Osteoblasts, inactive
- Osteoclasts
- Osteocytes
- Osteoid
- Periosteum
- Spicules
- Spongy bone
- Woven bone

Organ structures
- Articular cartilage
- Diaphysis
- Endosteum
- Epiphyseal plate
- Epiphysis
- Flat bone, diploe
- Flat bone, inner table
- Flat bone, outer table
- Hyaline cartilage
- Metaphysis
- Muscle
- Periosteum, osteogenic layer
- Spongy woven bone
- Suture

Deposition and resorption
- Cement lines

First lamellae
Haversian canal
Haversian canal contents
Inner circumferential lamellae
Interstitial lamellae
Osteon (Haversian system)
Outer circumferential lamellae
Resorption canal
Intramembranous formation
Canaliculi
Skeletal muscle
Skin
Spongy lamellar bone
Spongy woven bone

Endochondral formation
- Bone deposition
- Calcified cartilage
- Cartilage spicules
- Flat bones
- Long bones
- Periosteal band
- Resting zone
- Zone of degeneration
- Zone of ossification
- Zone of proliferation
- Zone of resorption
- Zones of maturation-hypertrophy-calcification

CHAPTER

6

Blood and Hematopoiesis

General Considerations

- In humans, the average blood volume is 5 liters, constituting 7% of the body mass.
- Blood is a specialized connective tissue consisting of cells and cell fragments (46% of blood volume) floating in a unique liquid extracellular matrix (54% of blood volume).
- Components
  - Cells and cell fragments
    - *Red blood cells (erythrocytes, RBCs)*, produced in the bone marrow
    - *White blood cells (leukocytes, WBCs)*, produced in the bone marrow; some lymphocytes are also produced in lymphoid tissues and organs.
    - *Platelets.* Cell fragments derived from *megakaryocytes* in the bone marrow; contain granules and function in blood coagulation; 150,000–450,000 per microliter blood
  - *Plasma.* Constitutes the extracellular matrix of blood
    - Composed of 90% water and 8–9% protein. Plasma proteins contains fibrinogen, a fiber precursor protein, which is converted into fibrin when blood clots.
    - *Serum.* Yellowish fluid remaining after blood has clotted.

Digital Histology: An Interactive CD Atlas with Review Text, by Alice S. Pakurar and John W. Bigbee
ISBN 0-471-64982-1

RED BLOOD CELLS

- Cells resemble bi-concave discs, 6–8 microns in diameter; 4–6 million per microliter of blood
- Cells are non-nucleated. Cytoplasm contains hemoglobin and cytoskeletal elements but lacks other organelles.
- Transport oxygen and carbon dioxide

WHITE BLOOD CELLS

- White blood cells are transported in the blood and migrate through vessel walls (diapedesis) to become active in connective tissues; 5–10 thousand per microliter of blood.
- *Granular leukocytes*
    - *Neutrophil (polymorphonuclear leukocyte, PMNs)*
        - 46–81% of circulating WBCs
        - Spherical cell, 12–15 microns in diameter; pale or unstained cytoplasmic granules; heterochromatic nucleus with three to five lobes
        - Move from the blood to sites of infection
        - Phagocytose bacteria and debris
    - *Eosinophil*
        - 1–3% of circulating WBCs
        - Spherical cell, 12–15 microns in diameter; cytoplasmic granules stain with eosin; bi-lobed nucleus
        - Move from the blood to sites of infection
        - Secrete proteins cytotoxic to parasites, neutralize histamine, and internalize antigen-antibody complexes
    - *Basophil*
        - <1% of circulating WBCs
        - Spherical cell, 12–15 microns in diameter; cytoplasmic granules stain dark blue with hematoxylin; S-shaped nucleus
        - Similar to mast cells; participate in the hypersensitivity reaction by secreting histamine and heparin
- *Agranular leukocytes*
    - *Lymphocyte*
        - 24–44% of circulating WBCs
        - Spherical cell, 6–8 microns in diameter; scant cytoplasm and a round heterochromatic nucleus often with a small indentation

- T and B lymphocytes
  - *T lymphocytes.* Originate in the bone marrow and mature in the thymus; provide cell-mediated immunity
  - *B lymphocytes.* Originate in the bone marrow and are carried in the blood to lymphoid tissues and organs, where they become activated and proliferate, transform into plasma cells in connective tissue, and provide humoral immunity by secreting antibodies
- *Monocyte*
  - 3–7% of circulating WBCs
  - Large spherical cells, 12–18 microns in diameter; abundant cytoplasm stains gray-blue; large, U-shaped, euchromatic nucleus
  - Enter connective tissue, where they transform into macrophages; function in phagocytosis and antigen presentation

HEMATOPOIESIS

➢ General considerations

- Hematopoiesis is the process of blood cell formation, beginning with a pleuripotential stem cell that subsequently goes through a series of cell divisions and differentiations to produce all the mature blood cells.
- Postnatal hematopoiesis occurs in red bone marrow located in the spongy bone region of long bones, vertebra, ribs, sternum, and skull. Lymphocytes are also generated in lymphoid organs and tissues.
- Blood cells have a relatively short life span and, therefore, new cells are formed continuously.
- Precursor cell lineage
  - *Stem cells.* Pleuripotential cells that give rise to all the blood cells; divide both to renew their own cell population as well as to form progenitor cells, thus beginning the process of blood cell formation.
  - *Progenitor cells.* Less potentiality than stem cells; committed to the formation of just one or two blood cell lines; have high mitotic activity, dividing to reproduce self and to from precursor cells.
  - *Precursor or blast cells.* Begin morphologic differentiation; display characteristics of the mature blood cells they will form; not self-renewing.

- *Mature blood cells.* Form after several cell divisions of the precursor or blast cells

➢ Erythropoiesis. Formation of erythrocytes

- Process that results in a non-nucleated cell filled with hemoglobin and specialized for transporting respiratory gases
- Stages. Cells listed in the order in which they form
    - *Proerythroblast.* Precursor cell
    - *Basophilic erythroblast.* Increased numbers of polyribosomes for hemoglobin production result in strong cytoplasmic basophilia; nucleus condenses.
    - *Polychromatophilic erythroblast.* Number of polyribosomes is reduced as hemoglobin accumulates in the cytoplasm.
    - *Orthochromatophilic erythroblast.* Continued condensation of the nucleus; increased eosinophilia of the cytoplasm due to accumulating hemoglobin
    - *Reticulocyte.* Extrusion of the nucleus; small number of polysomes remains.
    - *Mature erythrocyte*

➢ Granulopoiesis. Formation of granulocytes

- Process by which cells first produce lysosomal granules and then synthesize granules containing proteins specific for each granulocytic cell type. Although the term "granule" is commonly used to describe these structures, they are bounded by a unit membrane.
- Stages. Cells listed in the order in which they form
    - *Myeloblast.* Precursor cell
    - *Promyelocyte.* Production of azurophilic (blue) granules that contain lysosomal enzymes
    - *Myelocyte.* Nuclear condensation and the appearance of cell-specific granules containing proteins unique for each of the granular leukocytes
    - *Metamyelocyte.* Cell-specific granules continue to accumulate and the nucleus changes morphology to resemble that of the mature cell. Following the metamyelocyte, an additional stage, called a *band cell,* occurs in all granulocytes but is most prominent in the neutrophilic lineage.
    - *Mature neutrophils, eosinophils, and basophils*

➢ Monocytopoiesis. Formation of monocytes

- *Monoblast.* Precursor cell

- *Promonocyte*. Large cell, up to 18 microns in diameter; nucleus becomes indented and the cytoplasm is basophilic with numerous fine azurophilic granules (*lysosomes*).
- *Mature monocyte*

➢ Lymphocytopoiesis. Formation of lymphocytes

- *Lymphoblast*. Precursor cell
- *Pro-lymphocytes*. Reduction in size from lymphoblast; some remain in the bone marrow to produce B lymphocytes, others leave the bone marrow and travel to the thymus, where they complete their differentiation into T lymphocytes.

Structures Identified in this Section

Adipocytes
Basophil
Basophilic metamyelocyte
Bone marrow
Bone spicules
Diaphyseal bone
Eosinophil
Eosinophilic metamyelocyte
Erythrocyte
Hemopoietic bone marrow
Lymphocyte
Mature neutrophil
Medium lymphocyte
Megakaryocyte
Monocyte
Neutrophil
Neutrophilic band cell
Neutrophilic metamyelocyte
Orthochromatophilic erythroblast
Plasma
Plasma cell
Platelets
Polychromatophilic erythroblast
Segmenting neutrophil
Sinusoids

CHAPTER

7

MUSCLE TISSUE

GENERAL CONCEPTS

- Muscle tissue is specialized for the ability to shorten or contract. While all cells possess the cellular machinery necessary for shape change and contraction, these structures are significantly more prominent in muscle cells. For some muscle types, the cells are nonproliferative due to this high degree of specialization and differentiation.
- Muscle contraction is accomplished by the reciprocating sliding of intracellular filaments composed of actin and myosin.
- Muscle tissue comprises the "flesh" of the body and much of the walls of hollow organs. Due to its high degree of specialization, unique terms are used for certain structures in muscle cells.
    - Individual muscle cells are called *muscle fibers*.
    - The cytoplasm of muscle fibers is called *sarcoplasm*.
    - The muscle fiber plasma membrane is called the *sarcolemma*.
    - The smooth endoplasmic reticulum is called the *sarcoplasmic reticulum*.

Digital Histology: An Interactive CD Atlas with Review Text, by Alice S. Pakurar and John W. Bigbee
ISBN 0-471-64982-1

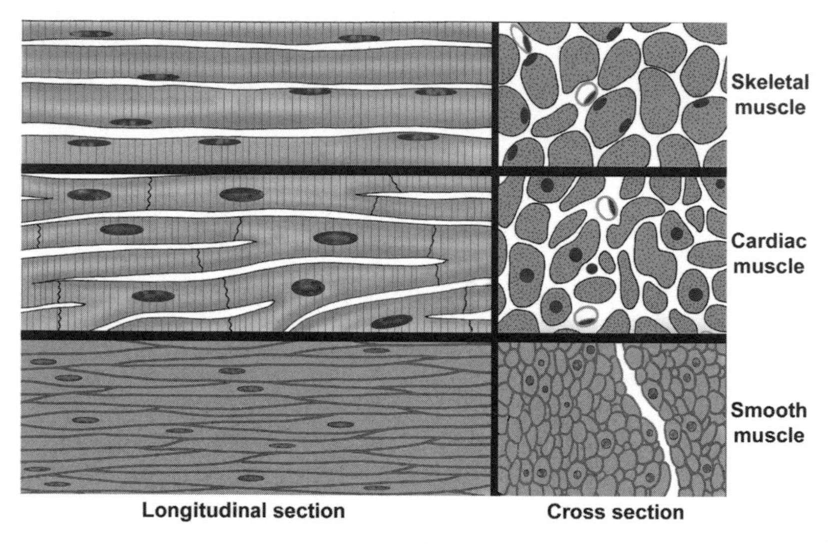

FIGURE 7.1. Comparison of muscle types.

CLASSIFICATION OF MUSCLE

- Functional classification is based on the type of neural control.
  - Voluntary
  - Involuntary
- Structural classification is based on the presence or absence of cross-striations.
  - Striated
  - Nonstriated (smooth)
- Combined functional and structural classification
  - *Skeletal muscle*
    - *Striated* and *voluntary*
    - Found mostly attached to the skeleton
  - *Cardiac muscle*
    - *Striated* and *involuntary*
    - Composes the majority of the heart wall (myocardium)
  - *Smooth (visceral) muscle*
    - *Nonstriated* and *involuntary*
    - Found mostly in the walls of hollow organs and vessels

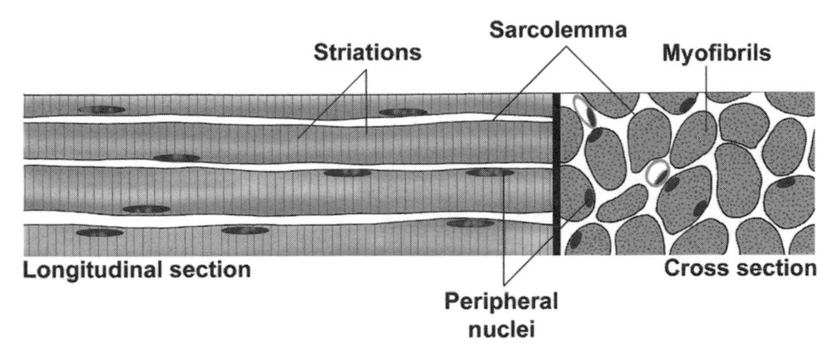

FIGURE 7.2. Skeletal muscle.

Skeletal Muscle

- Connective tissue investments of a skeletal muscle
  - Function
    - Separate muscle into compartments
    - Transmit the force of contraction to insertion points
  - Components
    - *Endomysium.* Reticular fibers surrounding each muscle fiber plus the external (basal) lamina produced by the muscle fiber
    - *Perimysium.* Dense connective tissue surrounding groups of fibers and dividing the muscle into fascicles
    - *Epimysium.* Dense connective tissue surrounding the entire muscle, blends with the deep fascia and tendons
- Hierarchy of skeletal muscle organization
  - *Myofilaments.* Visible only with the electron microscope; composed primarily of *actin,* which forms 5-nm wide *thin filaments,* and *myosin,* which forms 15-nm wide *thick filaments*
  - *Myofibrils.* Visible with the light microscope, 1–2 microns wide, oriented parallel to the long axis of the cell; composed of bundles of overlapping myofilaments that are arranged in register, producing an alternating light-dark, striated banding pattern
  - *Muscle fiber.* Specialized term for a muscle cell, 10–100 microns wide; sarcoplasm is filled with hundreds of myofibrils, which are oriented parallel to each other and to the long axis of the muscle fiber.
  - *Muscle fascicle.* Collection of muscle fibers surrounded by perimysium; collections of muscle fascicles are surrounded by the epimysium and form a muscle.

- Structure of skeletal muscle fibers
  - Largest fiber type, fibers can be 1–30 mm in length and 10–100 microns in diameter.
  - Each muscle fiber is cylindrical, unbranched, and multinucleated.
  - The multiple nuclei are located at the periphery of the muscle fiber immediately beneath the *sarcolemma*.
  - Extensive smooth endoplasmic reticulum is called the *sarcoplasmic reticulum*.
  - Each fiber is surrounded by an external lamina (basal lamina), which contributes to the endomysium of the muscle fiber.
  - Fibers can increase in size (hypertrophy) but not in number (hyperplasia).
  - Fibers show prominent, alternating light and dark bands (cross-striations) due to the alignment and overlap of the myofilaments within myofibrils. Myofilaments within a myofibril are arranged in register and adjacent myofibrils are similarly aligned, causing the banding pattern seen at both the light and electron microscopic levels.
    - *A band* appears dark and contains actin and myosin.
    - *I band* appears light and contains actin only.
    - *Z-line*, composed of alpha-actinin, is located in the center of the I band.
    - *H band* is located in the center of the A band and represents the area where actin is not present.
    - *M band* is located in the center of the H band and represents areas of cross-connections between myosin filaments.
    - *Sarcomere*
      - Contractile unit of striated muscle fibers, seen in both skeletal and cardiac muscle fibers
      - Extends from Z-line to Z-line
      - Sarcomeres are repeated in series along the length of each myofibril. Adjacent myofibrils maintain the alignment of sarcomeres.
    - Alterations in sarcomeres during contraction
      - Sarcomeres shorten.
      - Z-line interval narrows.
      - Width of H and I bands decrease as actin is pulled past the myosin.
      - A band width remains unchanged.

- Coordination of skeletal muscle fiber contraction
  - A complex system of intracellular, membranous structures called the triad insures coordinated contraction throughout the muscle fiber by (1) allowing the nervous impulse to penetrate and simultaneously reach all parts of the muscle fiber; and (2) releasing calcium in response to the nervous impulse. These functions are accomplished by the "triad."
  - *Triads.* Composed of one *T-tubule* plus two adjacent *terminal cisterns* of the sarcoplasmic reticulum
    - *T-tubules* are invaginations of the sarcolemma that occur at the junction between A and I bands of the myofibrils.
    - *Terminal cisterns* are expanded portions of the sarcoplasmic reticulum that lie adjacent to the T tubule and release calcium to initiate contraction.
  - Role of triad in muscle contraction
    - A nerve impulse arriving at the muscle fiber depolarizes the sarcolemma at the neuromuscular junction.
    - The membrane depolarization propagates along the sarcolemma and extends down the T-tubules.
    - T-tubule depolarization is transmitted to the terminal cisterns and the remainder of the sarcoplasmic reticulum, causing release of stored calcium.
    - Calcium initiates the interaction between actin and myosin myofilaments, leading to muscle contraction.
    - Calcium is recaptured by sarcoplasmic reticulum during relaxation.
- Mechanism of contraction, sliding filament model
  - At regions where actin and myosin myofilaments overlap, release of calcium causes the head groups of myosin to contact the actin filament.
  - Hydrolysis of ATP causes a change in the configuration of the myosin head group, resulting in a sliding of the actin myofilament past the myosin by the ratcheting action of the myosin head groups. Since the actin filaments are anchored at the Z-line, the result of the sliding is shortening of the sarcomere.
- Associated structures
  - *Neuromuscular junction (motor end plate)*
    - Specialized "synapse" between the terminals of a motor axon and the sarcolemma of a muscle fiber

- ◆ *Motor unit.* Consists of the motor neuron, its axon, and all the muscle fibers it innervates
- Proprioceptors
  - ◆ Sensory receptors, encapsulated by connective tissue, serve to regulate muscle tension and tone.
  - ◆ Types
    - ■ *Muscle spindle.* Highly modified skeletal muscle fibers, intrafusal fibers, are aligned with and surrounded by normal skeletal muscle fibers.
    - ■ *Golgi tendon organs.* Located within tendons

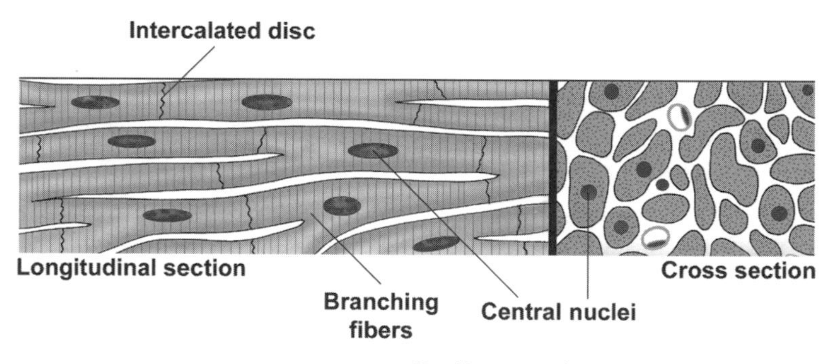

FIGURE 7.3. Cardiac muscle.

CARDIAC MUSCLE

➢ Cardiac muscle occurs only in the myocardium of the heart and, to a variable extent, in the roots of large vessels where they join the heart.

➢ Structure of cardiac muscle fibers

- Intermediate in size between skeletal and smooth muscle
- Fibers are cylindrical, branch, and form interwoven bundles.
- Usually one nucleus per fiber located in the center
- Organelles are clustered at the poles of the nucleus.
- Myofilament organization into myofibrils is identical to skeletal muscle. Cross-striations and bands identical to skeletal muscle are present, but not as prominent.
- *Intercalated discs*
  - ◆ Junctional complexes that are unique to cardiac muscle fibers

    - Consist of specialized cell junctions and interdigitations of the sarcolemma at the ends of the fibers
    - Contain three types of junctions
        - *Fascia adherens.* Similar to zonula adherens of epithelia; serve to attach cardiac muscle fibers and anchor actin filaments of the terminal sarcomeres at the ends of the cell. Acts as a hemi-Z-line.
        - *Desmosomes.* Bind ends of fibers together
        - *Gap junctions.* Provide ionic coupling between fibers
  - High vascularity and with large numbers of mitochondria reflect the high metabolic requirements of cardiac muscle fibers.
  - Fibers are capable of hypertrophy but not hyperplasia.
- Coordination of cardiac muscle contraction
  - Sarcomeres, myofibrils, and myofilaments are the same as skeletal muscle fibers.
  - T-tubules are located at the level of the Z-lines, rather than at junction of A and I bands as in skeletal muscle.
  - No triads. Sarcoplasmic reticulum is not as well developed as in skeletal muscle fibers and does not form terminal cisterns. Contraction is initiated by intracellular calcium release.
  - Contraction can spread through the myocardium due to the presence of gap junctions that allow calcium to flow from one fiber into another.

Smooth Muscle

- Smooth muscle occurs mostly as sheets, which form the walls of most hollow organs with the exception of the heart. Smooth muscle is also

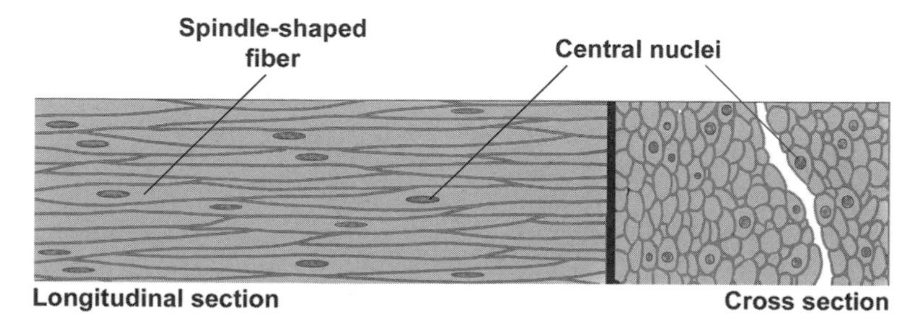

FIGURE 7.4. Smooth muscle.

prominent in the walls of blood vessels, many respiratory passageways, and some genital ducts.

- Structure of smooth muscle fibers
  - Smallest fiber type, length varies from 20 microns in blood vessels to 500 microns in the uterus
  - Unbranched spindle-shaped fibers are elongated with tapering ends and unbranched.
  - Possess a single, centrally placed, oval nucleus, which can appear spiraled or "inch-worm"–shaped when the fiber is contracted.
  - Organelles are clustered at the poles of the nucleus.
  - Nonstriated; no myofibrils are present.
  - External (basal) lamina is present along with reticular fibers.
  - Abundant gap junctions
  - Capable of both hypertrophy and hyperplasia
- Organization of the contractile proteins
  - Actin and myosin myofilaments are present, but they are not organized into myofibrils.
  - Myofilaments overlap as in striated muscle and crisscross throughout the sarcoplasm, forming a reticulum.
  - *Dense bodies*
    - Serve as insertion points for myofilaments to transmit the force of filament sliding
    - Contain alpha-actinin and, thus, resemble Z-lines of striated muscle
    - Present in the cytoplasm and associated with the sarcolemma
- Coordination of smooth muscle contraction
  - No T-tubules are present; however, fibers do have a rudimentary sarcoplasmic reticulum.
  - Sliding filament mechanism. Regulated by intracellular release of calcium but with some differences from striated muscle fibers
- Types of smooth muscle
  - *Visceral smooth muscle*
    - Occurs in sheets in the wall of hollow organs (e.g., digestive tract)
    - Minimally innervated; contraction spreads in peristaltic waves facilitated by large numbers of gap junctions.
    - Specialized for slow, prolonged contraction

- *Multiunit smooth muscle*
  - Richly innervated
  - Specialized for precise, graded contraction (e.g., iris of the eye)

Structures Identified in This Section

- Skeletal muscle
  - Connective tissue components of skeletal muscle
    - Endomysium
    - Epimysium
    - Fascicle
    - Perimysium
  - Skeletal muscle fiber
    - A band
    - Cross-striations
    - I band
    - Myofibrils
    - Nucleus
    - Sarcolemma
    - Sarcomere
    - Z-line
  - Motor end plate
    - Myelinated axons
    - Presynaptic terminals
  - Proprioceptors
    - Extrafusal muscle fibers
    - Intrafusal muscle fibers
    - Muscle spindle
    - Sensory axons
- Cardiac muscle
  - Cardiac muscle fibers
    - A band
    - Cross-striations
    - Glycogen granules
    - H band
    - I band
    - Intercalated disc
    - M band
    - Mitochondria
    - Myofibrils
    - Myofilaments
    - Nucleus
    - Sarcomere
    - Z-lines
  - Capillaries
  - Connective tissue
- Smooth muscle
  - Autonomic ganglion
  - Small intestine
    - Inner circular layer of smooth muscle
    - Outer longitudinal layer of smooth muscle
    - Mucosa
  - Smooth muscle fascicle
  - Smooth muscle fibers
    - Nucleus
    - Sarcolemma
    - Spindle shape
    - Autonomic neuron cell body

CHAPTER

8

NERVOUS TISSUE

GENERAL CONSIDERATIONS

➢ Nervous tissue is highly specialized to employ modifications in membrane electrical potentials to relay signals throughout the body. Neurons form intricate circuits that (1) relay sensory information from the internal and external environments; (2) integrate information among millions of neurons; and (3) transmit effector signals to muscles and glands.

➢ Anatomical subdivisions of nervous tissue
- *Central nervous system (CNS)*
  - *Brain*
  - *Spinal cord*
- *Peripheral nervous system (PNS)*
  - *Nerves*
  - *Ganglia (singular, ganglion)*

CELLS OF NERVOUS TISSUE

➢ *Neurons*
- Functional units of the nervous system; receive, process, store, and transmit information to and from other neurons, muscle cells, or glands

Digital Histology: An Interactive CD Atlas with Review Text, by Alice S. Pakurar and John W. Bigbee
ISBN 0-471-64982-1

- Composed of a cell body, dendrites, axon and its terminal arborization, and synapses
- Form complex and highly integrated circuits

➢ *Supportive cells*

- Outnumber neurons 10:1
- Provide metabolic and structural support for neurons, insulation (myelin sheath), homeostasis, and phagocytic functions
- Comprised of astrocytes, oligodendrocytes, microglia, and ependymal cells in the CNS; comprised of Schwann cells in the PNS

STRUCTURE OF A "TYPICAL" NEURON

➢ *Cell body (soma, perikaryon)*

- *Nucleus*
  - ◆ Large, spherical, usually centrally located in the soma
  - ◆ Highly euchromatic with a large, prominent nucleolus
- Cytoplasm
  - ◆ Well-developed cytoskeleton
    - ■ Intermediate filaments (*neurofilaments*). 8–10 nm in diameter
    - ■ Microtubules. 18–20 nm in diameter
  - ◆ Abundant rough endoplasmic reticulum and polysomes (*Nissl substance*)
  - ◆ Well-developed Golgi apparatus
  - ◆ Numerous mitochondria

➢ *Dendrite(s)*

- Usually multiple and highly branched at acute angles
- May possess *spines* to increase surface area for synaptic contact
- Collectively, form the majority of the receptive field of a neuron; conduct impulses toward the cell body
- Organelles
  - ◆ Microtubules and neurofilaments
  - ◆ Rough endoplasmic reticulum and polysomes
  - ◆ Smooth endoplasmic reticulum
  - ◆ Mitochondria

➢ *Axon*

- Usually only one per neuron

- Generally of smaller caliber and longer than dendrites
- Branches at right angles, fewer branches than dendrites
- Organelles
  - Microtubules and neurofilaments
  - Lacks rough endoplasmic reticulum and polysomes
  - Smooth endoplasmic reticulum
  - Mitochondria
- *Axon hillock.* Region of the cell body where axon originates
  - Devoid of rough endoplasmic reticulum
  - Continuous with *initial segment* of the axon that is a highly electrically excitable zone for initiation of nervous impulse
- Usually ensheathed by supporting cells
- Transmits impulses away from the cell body to
  - Neurons
  - Effector structures. Muscle and glands
- Terminates in a swelling, the terminal bouton, which is the presynaptic element of a synapse

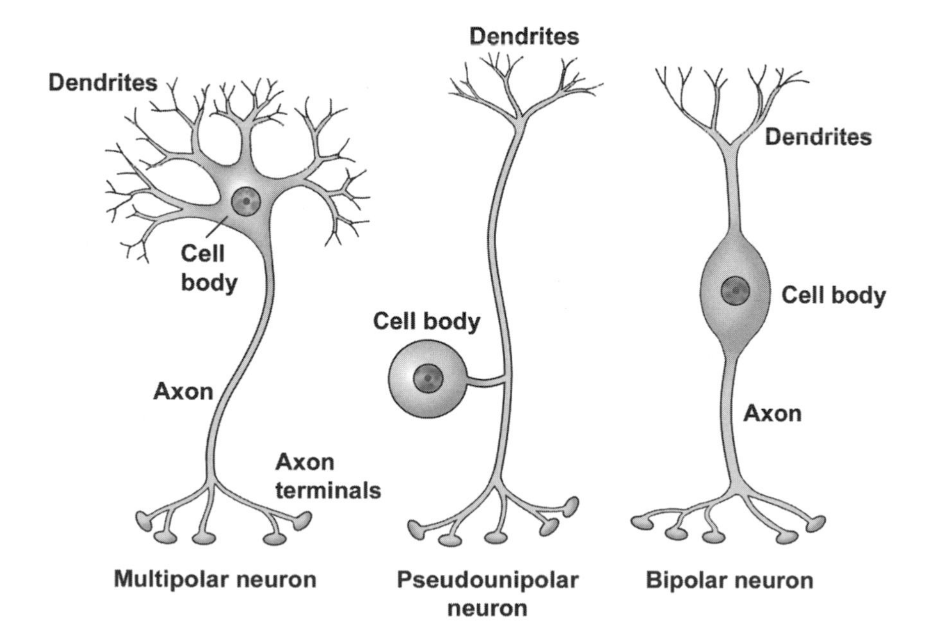

FIGURE 8.1. Types of neurons based on shape.

Type of Neurons by Shape and Function

- *Multipolar neuron.* Most numerous and structurally diverse type
  - Efferent. Motor or integrative function
  - Found throughout the CNS and in autonomic ganglia in the PNS
- *Pseudounipolar neuron*
  - Afferent. Sensory function
  - Found in selected areas of the CNS and in sensory ganglia of cranial nerves and spinal nerves (dorsal root ganglia)
- *Bipolar neuron*
  - Afferent. Sensory function
  - Found associated with organs of special sense (retina of the eye, olfactory epithelium, vestibular and cochlear ganglia of the inner ear)
  - Developmental stage for all neurons

Arrangement of Neuronal Cell Bodies and Their Processes

- In both CNS and PNS, cell bodies are found in clusters or layers and axons travel in bundles. These groupings are based on common functions and/or common connections.

	Group of cell bodies	Bundle of processes
Central nervous system	Nucleus or cortex (gray matter)	Tract (white matter)
Peripheral nervous system	Ganglion	Nerve

Synapse

- The function of the synapse is to alter the membrane potential of the postsynaptic target cell to either facilitate or inhibit the likelihood of the stimulus to be propagated by the postsynaptic cell. Most neurons receive thousands of synaptic contacts, both stimulatory and inhibitory, and the algebraic sum of these inputs determines whether the postsynaptic cell will depolarize.
- Classified according to postsynaptic target
  - *Axodendritic.* Most common
  - *Axosomatic*

- *Axoaxonic.* Mostly occur at presynaptic terminals
- *Neuromuscular junction*

➢ Structure of the synapse

- *Presynaptic component*
  - Distal end of the axon branches, each branch terminating in a swelling or button called the *terminal bouton.*
  - Bouton contains *synaptic vesicles/granules,* which contain neurotransmitters and numerous mitochondria.
- *Synaptic gap/cleft.* Separation (20–30 nm) between pre- and postsynaptic cells.
- *Postsynaptic component*
  - Formed by the membrane of the postsynaptic neuron or muscle cell and contains receptors for neurotransmitters
  - Membrane shows a *postsynaptic density* or thickening on its cytoplasmic side.
- *Bouton en passant.* "Bouton like" swellings along the length of an axon, allows a single axon to contact many distant cells. Common in smooth muscle innervation.

THE REFLEX ARC

➢ The reflex arc is the simplest neuronal circuit and includes each of the elements discussed above. These circuits provide rapid, stereotyped reactions to help maintain homeostasis. To begin the reflex, a pseudounipolar, sensory neuron is activated by a receptor. The axon carries an afferent signal from the skin into the spinal cord where it synapses on a multipolar association neuron or interneuron. The interneuron signals a multipolar, motor neuron whose axon then carries an efferent signal to skeletal muscle to initiate contraction.

SUPPORTIVE CELLS

➢ Supporting cells of the CNS (neuroglial cells); outnumber neurons 10:1

- *Astrocytes*
  - Stellate morphology
  - Types
    - *Fibrous astrocytes* in white matter
    - *Protoplasmic astrocytes* in gray matter

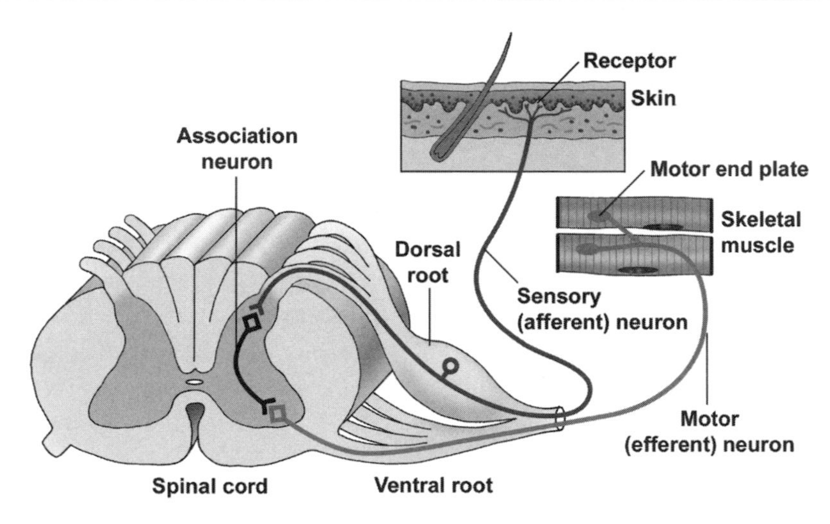

FIGURE 8.2. The reflex arc.

- ◆ Functions
  - ■ Physical support
  - ■ Transport nutrients
  - ■ Maintain ionic homeostasis
  - ■ Take up neurotransmitters
  - ■ Form glial scars (gliosis)
- *Oligodendrocytes*
  - ◆ Present in white and gray matter
  - ◆ *Interfascicular oligodendrocytes* are located in the white matter of the CNS, where they produce the myelin sheath.
- *Ependymal cells.* Line ventricles
- *Microglia*
  - ◆ Not a true neuroglial cell; derived from mesoderm whereas neuroglial cells, as well as neurons, are derived from ectoderm
  - ◆ Highly phagocytic cells

➢ Supporting cells of the PNS. *Schwann cells*

- *Satellite Schwann cells* surround cell bodies in ganglia

- *Ensheathing Schwann cells*
    - Surround unmyelinated axons. Numerous axons indent the Schwann cell cytoplasm and are ensheathed only by a single wrapping of plasma membrane.
    - Produce the myelin sheath around axons

Myelin Sheath

- The *myelin sheath* is formed by the plasma membrane of supporting cells wrapping around the axon. The sheath consists of multilamellar, lipid-rich segments produced by Schwann cells in the PNS and oligodendrocytes in the CNS.
- Functions
    - Increases speed of conduction (*saltatory conduction*)
    - Insulates the axon
- Similar structure in CNS and PNS with some differences in protein composition
- Organization
    - *Internode.* Single myelin segment
    - *Paranode.* Ends of each internode where they attach to the axon
    - *Node of Ranvier.* Specialized region of the axon between myelin internodes where depolarization occurs
- In the PNS, each Schwann cell associates with only one axon and forms a single internode of myelin.
- In the CNS, each oligodendrocyte associates with many (40–50) axons (i.e. each oligodendrocyte forms multiple internodes on different axons).

Connective Tissue Investments of Nervous Tissue

- Peripheral nervous system
    - *Endoneurium.* Delicate connective tissue surrounding Schwann cells; includes the basal lamina secreted by Schwann cells as well as reticular fibers
    - *Perineurium.* Dense tissue surrounding groups of axons and their surrounding Schwann cells, forming fascicles; forms the blood-nerve barrier
    - *Epineurium.* Dense connective tissue surrounding fascicles and the entire nerve

- Central nervous system
  - *Meninges*
    - *Pia mater*
      - Thin membrane lying directly on the surface of the brain and spinal cord
      - Accompanies larger blood vessels into the brain and spinal cord
    - *Arachnoid membrane*
      - Separated from pia mater by connective tissue trabeculae
      - Encloses the *subarachnoid space,* which contains blood vessels and the *cerebrospinal fluid (CSF)* produced by the cells of the choroid plexus
      - Together with pia mater, constitute the *leptomeninges;* inflammation of these membranes produces meningitis
    - *Dura mater*
      - Outermost of the meninges
      - Dense connective tissue that includes the periosteum of the skull

STRUCTURES IDENTIFIED IN THIS SECTION

Glial cells
  Astrocyte, protoplasmic
  Astrocyte, fibrous
  Astrocyte nuclei
  Astrocytic end feet
  Microglial cell nuclei
  Myelin sheath
  Oligodendrocyte nuclei
  Oligodendrocyte, satellite
  Oligodendrocyte, interfascicular
Grey matter
Meninges
  Arachnoid
  Dura mater
  Pia mater
  Subarachnoid space
  Subdural space

Neuron Types
  Bipolar neurons
    Central axons
    Peripheral axons
    Cochlear branch of cranial nerve VIII
    Organ of Corti
    Bone
  Multipolar neurons
    Axon
    Axon hillock
    Cell body
    Dendrite
    Nissl substance
    Nucleolus
    Nucleus
    Types

        - Autonomic ganglion
        - Purkinje cell (neuron)
            - Purkinje cell body
            - Purkinje cell dendrites
            - Dendritic spines
        - Pyramidal neuron
            - Apical dendrites
    - Pseudounipolar neurons
        - Axons
        - Dorsal root ganglion
        - Myelin
        - Satellite Schwann cells
- Peripheral nerve
    - Adipose tissue
    - Axon
    - Basal lamina
    - Blood vessels
    - Connective tissue
    - Duct of sweat glands
    - Endoneurium
    - Epineurium
    - Microtubules
    - Muscle tissue
    - Myelin lamella
    - Myelin sheath
    - Nerve fascicle
    - Neurofilaments
    - Node of Ranvier
    - Paranodal loops
    - Paranodal region
    - Perineurium
    - Schwann cell nucleus
    - Schwann cell process
    - Unmyelinated axons
- Receptors
    - Axon
    - Meissner's corpuscle
    - Muscle spindle
        - Skeletal muscle fibers
        - Modified skeletal muscle fibers
        - Capsule
        - Sensory axon
    - Pacinian corpuscle
        - Perineurial cells
- Spinal cord
- Spinal nerve roots
- Synapses
    - Motor end plate
        - Skeletal muscle
        - Axons
    - CNS synapse
        - Terminal bouton
        - Synaptic vesicles (Neurotransmitter vesicles)
        - Mitochondria
        - Synaptic cleft
        - Postsynaptic cell
        - Postsynaptic density
        - Dendrite
        - Dendritic spine
- White matter

CHAPTER 9

CONCEPTS AND TERMINOLOGY

LUMEN

➢ The cavity or channel within a hollow organ.

MEMBRANES

➢ Definition. A layer of epithelium and its underlying connective tissue that covers a surface of the body. A membrane lines almost all surfaces of the body.

➢ Types of membranes
- *Cutaneous membrane* or *skin*
  - Covers the exterior surface of the body.
  - Composition
    - Stratified squamous keratinized epithelium called epidermis
    - Two layers of connective tissue, loose connective tissue and dense irregular connective tissue, called the dermis
  - May possess hair follicles, sweat glands, and sebaceous glands
- *Mucous membrane* or *mucosa*
  - Lines all interior lumens of organs that open to the exterior, such as stomach, uterus, and trachea.

Digital Histology: An Interactive CD Atlas with Review Text, by Alice S. Pakurar and John W. Bigbee
ISBN 0-471-64982-1

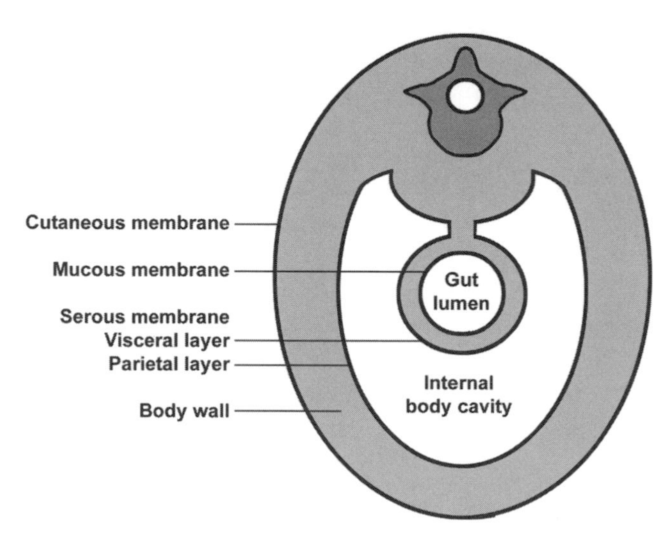

FIGURE 9.1. Cross-section through the abdomen, illustrating epithelial membranes.

- Composition
  - Epithelium varies depending on the location.
  - The lamina propria, composed of loose connective tissue, lies beneath the epithelium.
  - Muscularis mucosae, a layer of smooth muscle, is frequently, but not always, present in a mucous membrane.
- Many mucous membranes are associated with mucus-secreting cells or glands that lubricate the surface of the membrane.

- *Serous membrane* or *serosa*
  - Location
    - Lines the perimeter of all internal body cavities that do not open to the exterior, such as peritoneal cavity (called peritoneum), pleural cavity (called pleura), and pericardial cavity (called pericardium). This lining constitutes the parietal layer of these cavities.
    - Covers and forms the outer layer of any organs that protrude into those body cavities, such as stomach and jejunum (peritoneum), lungs (pleura), and heart (pericardium). This covering constitutes the visceral layer of these cavities.
  - Composition
    - Simple squamous epithelium called mesothelium
    - Loose connective tissue

Cortex vs. Medulla

- Components
  - *Cortex.* The outer region or portion of some organs, such as kidney, lymph nodes, and adrenal glands; surrounds the internal medulla
  - *Medulla.* The center portion of some organs; surrounded by a cortex
- Differentiation of these two subdivisions is a result of their different components, origins and/or functions.

Stroma vs. Parenchyma

- *Stroma.* The supporting framework of an organ, usually composed of connective tissue
- *Parenchyma.* The cells and tissues of an organ that perform the function of the organ; composed of epithelium, muscle, nerve and, sometimes, connective tissue

Structures Identified in This Section

Lumen

Membranes
- Adventitia
- Connective tissue of serosa
- Cutaneous membrane
- Dermis
- Epidermis
- Epithelium
- Keratin
- Lamina propria
- Lumen
- Mesothelium
- Mucosa
- Mucous membrane
- Muscularis externa
- Muscularis mucosae
- Peritoneal space
- Serosa
- Serous membrane
- Simple squamous epithelium
- Submucosa

Cortex and medulla
- Cortex
- Medulla

Stroma and parenchyma
- Parenchyma
- Stroma

CHAPTER

10

Cardiovascular System

General Considerations

- Continuous tubular system for transporting blood, carrying oxygen, carbon dioxide, hormones, nutrients, and wastes
- Components of the circulatory system
    - *Heart*. Highly modified, muscular blood vessel specialized for pumping the blood. Composed of two atria and two ventricles.
    - *Closed circuit of vessels*. The vessels are listed below in the order that blood would follow as it leaves the heart.
        - *Elastic arteries* (e.g., aorta and pulmonary arteries)
        - *Muscular arteries* (remaining named arteries)
        - *Small arteries* and *arterioles*
        - *Capillaries*
        - *Venules* and *small veins*
        - *Medium veins* (most named veins)
        - *Large veins* (e.g., venae cavae return blood to the heart)
- Circuitry of the circulatory system
    - *Pulmonary circulation*
        - Circuit of blood between the heart and lungs

Digital Histology: An Interactive CD Atlas with Review Text, by Alice S. Pakurar and John W. Bigbee
ISBN 0-471-64982-1

- Blood leaves the right ventricle of the heart through *pulmonary arteries* and proceeds through a series of smaller arteries to supply *pulmonary capillaries* in the lungs. Blood returns through a series of increasingly larger veins to the *pulmonary veins* to the left atrium.
- Functions for exchange of carbon dioxide and oxygen between the blood and atmosphere

- *Systemic circulation*
  - Circuit that distributes blood from the heart to the body tissues
  - Blood leaves the left ventricle of the heart through the *aorta* and proceeds through a series of smaller arteries to supply *systemic capillaries* throughout the body. Blood returns through a series of increasingly larger veins via the *superior* and *inferior venae cavae* to the right atrium.
  - Functions for exchange of carbon dioxide and oxygen, and nutrients and metabolic wastes between the blood and tissues; distribution of hormones.
- *Lymphatic circulation.* Consists of a system of blind-ended lymph vessels positioned throughout the body, which return tissue fluid to the venous circulation.

Basic Structural Organization

- The walls of the entire cardiovascular system, consists of three concentric layers or tunics that are continuous between both the heart and vessels. The constituents and thickness of these layers vary depending on the mechanical and metabolic functions of the vessel.
- *Inner tunic*
  - In the heart, this layer is called the *endocardium*; in vessels it is termed the *tunica intima*.
  - Composition
    - Simple squamous epithelium (*endothelium*)
    - Varying amounts and types of connective tissue
    - In the largest vessels, longitudinally oriented smooth muscle may be present in the connective tissue layer.
- *Middle tunic*
  - In the heart this layer is composed of cardiac muscle and is called the *myocardium*.

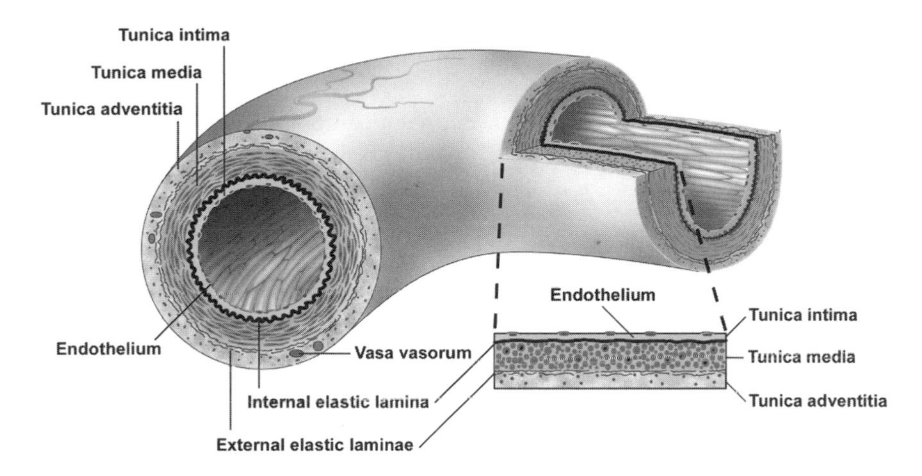

FIGURE 10.1. Structure of a muscular artery.

- In vessels this layer is composed of circularly oriented smooth muscle or smooth muscle plus connective tissue and is called the *tunica media.*

➢ *Outer tunic*

- In the heart, this layer consists of a serous membrane, called the *epicardium (visceral pericardium)* composed of connective tissue covered with a simple squamous epithelium (mesothelium).
- In vessels, this layer is called the *tunica adventitia* and is composed of connective tissue; variable amount of longitudinally arranged smooth muscle is present in this layer in the largest veins.
- Possesses blood vessels that supply the wall of the heart or larger blood vessels
  - ◆ *Coronary blood vessels.* Supply the heart wall
  - ◆ *Vasa vasorum.* Consists of a system of small blood vessels that supply the outer wall of larger vessels

ARTERIES

➢ General considerations

- Carry blood away from the heart and toward capillary beds
- Have thicker walls and smaller lumens than veins of similar size
- Tunica media is the predominate tunic.
- Cross-sectional outlines are more circular in arteries than in veins.

- Types
  - *Elastic (large) arteries (aorta, pulmonary arteries)*
    - Internal elastic lamina is present but difficult to distinguish.
    - Tunica media is composed of fenestrated sheets of elastic tissue (*elastic lamellae*) and smooth muscle.
    - Passively maintain blood pressure by distension and recoil of the elastic sheets
  - *Muscular (medium, distributing)*
    - Tunica media is composed of smooth muscle.
    - *Internal elastic lamina.* Single, fenestrated, elastic sheet; lies internal to the smooth muscle of the tunica media.
    - *External elastic laminae.* Multiple elastic sheets; lie external to the smooth muscle of the tunica media
    - Regulate blood pressure and blood distribution by contraction and relaxation of smooth muscle in the tunica media
  - *Small arteries and arterioles*
    - Less than 200 microns in diameter
    - Small arteries have an internal elastic lamina and up to eight layers of smooth muscle in the tunica media.
    - Arterioles usually lack an internal elastic lamina and have one to two layers of smooth muscle in the tunica media.
    - Arterioles are the vessels that regulate blood pressure and deliver blood under low pressure to capillaries.

CAPILLARIES

- General considerations
  - Function to exchange oxygen and carbon dioxide and nutrients and metabolic wastes between blood and cells
  - Lumen is approximately 8 microns in diameter, thus only large enough for RBCs to move through in a single row.
  - Composed of the endothelium (simple squamous epithelium) and its underlying basal lamina
- Types
  - *Continuous capillaries*
    - Most common
    - Endothelium is continuous (i.e., has no pores)

- *Fenestrated capillaries*
  - Endothelium contains pores that may or may not be spanned by a *diaphragm*. If present, the diaphragm is thinner than two apposed plasma membranes.
  - Pores with diaphragms are common in capillaries in the endocrine organs and portions of the digestive tract. Pores lacking diaphragms are uniquely present in the glomerular capillaries of the kidney.
  - Pores facilitate diffusion across the endothelium
- *Discontinuous sinusoidal capillaries*
  - Larger diameter and slower blood flow than in other capillaries
  - Endothelium has large pores that are not closed by a diaphragm.
  - Gaps are present between adjacent endothelial cells.
  - Partial or no basal lamina present.
  - Prominent in spleen and liver

VEINS

➢ General considerations

- Return blood from capillary beds to the heart
- Have thinner walls and larger lumens than arteries of similar size; cross-sectional outlines are more irregular
- Tunica adventitia is the predominate tunic.
- Larger veins possess valves, that are extensions of the tunica intima that serve to prevent back-flow of blood.
- Types
  - *Venules* and *small veins*
    - Tunica media is absent in venules. Smooth muscle fibers appear in the tunica media as venules progress to small veins.
    - *High endothelial venules*. Venules in which the endothelium is simple cuboidal; facilitate movement of cells from the blood into the surrounding tissues (diapedesis). This type of venule is found in many of the lymphatic tissues.
  - *Medium veins*. Smooth muscle forms a more definitive and continuous tunica media; most named veins are in this category.
  - *Large veins*, includes superior and inferior venae cavae; have well-developed, longitudinally oriented smooth muscle in the tunica adventitia in addition to the smooth muscle in the tunica media.

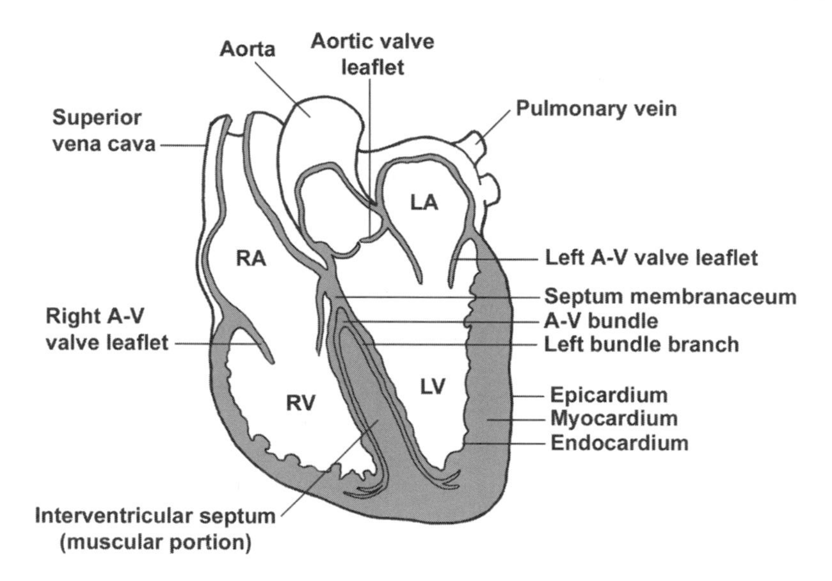

FIGURE 10.2. Diagram of a frontal section of the heart (RA, right atrium; LA, left atrium; RV, right ventricle; LV, left ventricle).

HEART

- Develops by a vessel folding back on itself to produce four chambers in the adult. Two upper chambers, *atria* (singular, *atrium*), receive blood from the body and lungs; two *ventricles* pump blood out of the heart.
- Impulse conducting system. Formed of specialized cardiac muscle fibers that initiate and coordinate the contraction of the heart
  - *Sinoatrial (SA) node* in the right atrium is the electrical *pacemaker* that initiates the impuse.
  - Fibers spread the impulse throughout the atria as well as transferring it to the atrioventricular node.
  - The *atrioventricular (AV) node* is located in the *interatrial septum.*
  - An *atrioventricular bundle* extends from the AV node in the *septum membranaceum* and bifurcates into right and left *bundle branches* that lie beneath the endocardium on both sides of the *interventricular septum.*
  - *Purkinje fibers*, modified, enlarged cardiac muscle fibers leave the bundle branches to innervate the myocardium.
- Tunics
  - *Endocardium*
    - Homologous to the tunica intima of vessels

    - Consists of an endothelium (simple squamous epithelium) plus underlying connective tissue
    - *Cardiac valves.* Folds of the endocardium
        - *Semilunar valves* at the base of the aortic and pulmonary trunks prevent backflow of blood into the heart.
        - *Atrioventricular valves (bicuspid and tricuspid)* prevent backflow of blood from the ventricles into the atria.
- *Myocardium*
    - Composed of cardiac muscle
    - Fibers insert on components of the cardiac skeleton.
    - Thickest layer of the heart
    - Variation in thickness depends on the function of each chamber; thicker in ventricles than atria and thicker in left ventricle than right ventricle
- *Epicardium (visceral pericardium)*
    - Serous membrane on the surface of the myocardium
    - Consists of a simple squamous epithelium and a loose connective tissue, with adipocytes, adjacent to the myocardium.
    - Coronary blood vessels are located in the connective tissue.
- *Cardiac skeleton.* Thickened regions of dense connective tissue that provide support for heart valves and serve as insertion of cardiac muscle fibers
    - *Annuli fibrosi* are connective tissue rings that surround and stabilize each valve.
    - *Septum membranceum* is a connective tissue partition forming the upper portion of the interventricular septum; this connective tissue also separates the left ventricle from the right atrium.

Structures Identified in This Section

Blood cells
- Neutrophil (polymorphonuclear, PMN)
- Eosinophil
- Basophil
- Lymphocyte
- Monocyte
- Platelets

Heart
- Cardiac skeleton and valves
    - Annulus fibrosus
    - Aortic (seminlunar) valve
    - Atrioventricular valve
    - Septum membranaceum
- Chambers
    - Left atrium

- Left ventricle
- Right atrium
- Right ventricle
- Conducting system
  - Atrioventricular bundle (of His)
  - Purkinje fibers
  - Right and left bundle branches
- Wall
  - Cardiac muscle fibers
  - Endocardium
  - Epicardium
  - Interatrial septum
  - Intercalated discs
  - Interventricular septum
  - Myocardium

Vessels
- Structures
  - Elastic lamellae
  - Endothelium
  - External elastic laminae
  - Internal elastic lamina
  - Pericyte
  - Smooth muscle cells (longitudinal and circular)
  - Subendothelium of tunica intima
  - Tunica adventitia
  - Tunica intima
  - Tunica media
  - Vasa vasorum
- Types
  - Arteriole
  - Elastic artery (aorta)
  - High endothelial venule
  - Large vein (inferior vena cava)
  - Muscular artery
  - Small/medium vein
  - Transitional artery
  - Venule
- Capillaries (blood)
  - Continuous
  - Fenestrated (with endothelial cell fenestrae and diaphragms)
  - Discontinuous sinusoidal capillary
- Lymphatic capillaries
- Lymphatic vessels

CHAPTER

11

SKIN

FUNCTIONS OF SKIN

- Protection against physical abrasion, chemical irritants, pathogens, UV radiation, and dessication
- Thermoregulation
- Reception of pressure and touch sensations
- Production of vitamin D
- Excretion

COMPONENTS OF THE INTEGUMENT

- *Epidermis*. Stratified squamous keratinized epithelium
- *Dermis*. Composed of two layers of connective tissue containing blood vessels, nerves, sensory receptors, and sweat and sebaceous glands. Beneath the dermis is a layer of loose connective and adipose tissues that forms the superficial fascia of gross anatomy termed the *hypodermis*. This layer is considered along with the skin, though technically it is not part of the integument.

Digital Histology: An Interactive CD Atlas with Review Text, by Alice S. Pakurar and John W. Bigbee
ISBN 0-471-64982-1

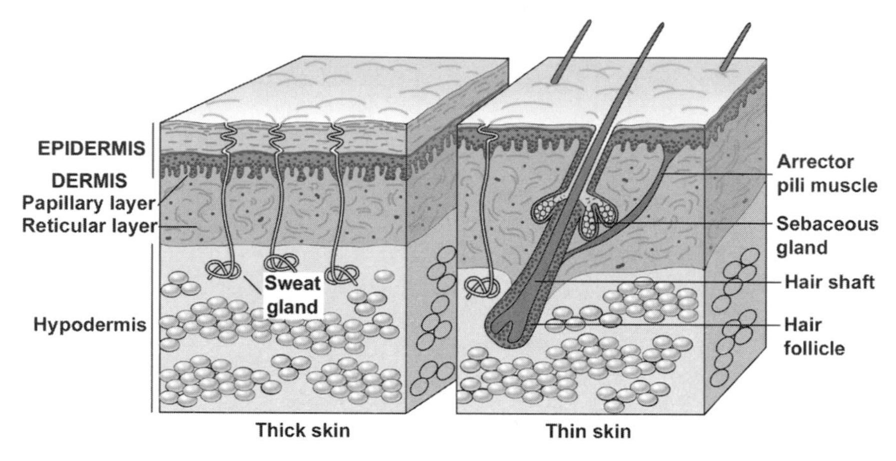

FIGURE 11.1. Structure of thin and thick skin.

CLASSIFICATION OF SKIN—BASED ON THE THICKNESS OF THE EPIDERMIS

- *Thin skin*
  - Covers entire body except palms and soles; 0.5 mm thick on the eyelid, 5 mm thick on the back and shoulders
  - Epidermis is thin, 0.075–0.15 mm thick, but the dermis can be quite thick.
  - Possesses hair with sebaceous glands
  - Sweat glands are present.
- *Thick (glabrous) skin*
  - Located on palms of the hands and soles of the feet; 0.8–1.5 mm thick
  - Epidermis is 0.4–0.6 mm thick.
  - Hairless and, thus, possesses no sebaceous glands
  - Sweat glands are present.

EPIDERMIS

- Cell types
  - *Keratinocytes*. Keratinizing epidermal cells, major cell type in the epidermis
  - *Melanocytes*. Melanin pigment-producing cells
  - *Langerhans cells*. Macrophages, antigen-presenting cells

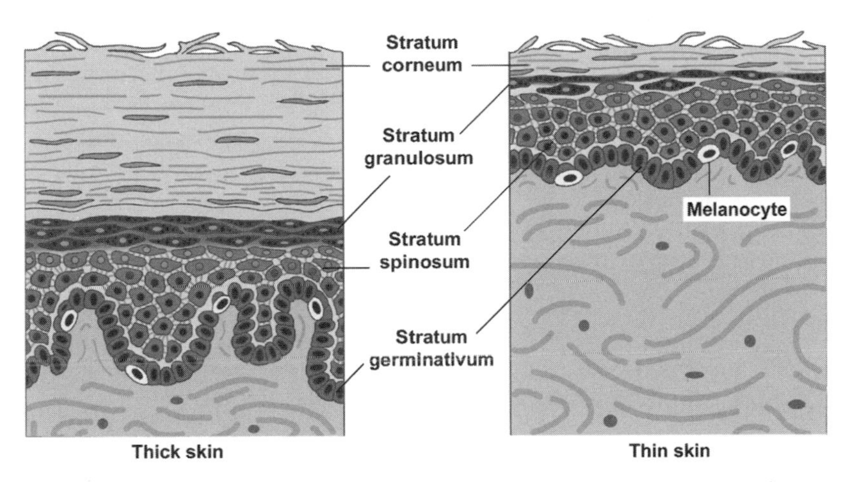

FIGURE 11.2. Comparison of epidermal layers in thick and thin skin.

➢ Layers of the epidermis and keratinization

- The epidermis is a stratified squamous, keratinized (dry) epithelium. It is continually renewed every 15–30 days. Rapid cell proliferation occurs in the deepest layer (stratum germinativum) and daughter cells differentiate as they migrate toward the surface. This differentiation involves a process called keratinization, which results in a variably thick layer of nonliving cellular husks at the surface of the epidermis. All cells in the epidermis that undergo the keratinization process are called keratinocytes.
- Layers of the epidermis
  - *Stratum germinativum*—(basale). A single layer of cuboidal to columnar shaped cells that rest on the basement membrane and undergo rapid cell proliferation. These cells contain intermediate filaments composed of keratin.
  - *Stratum spinosum*. "Prickle-cell" or spiny cell layer; 3–10 cells thick. This layer is so-called because the cells are attached to one another by desmosomes, and the cellular shrinkage resulting from fixation produces the spine-like structures. These cells accumulate bundles of keratin filaments called *tonofibrils*.
  - *Stratum granulosum*: two to four cells thick; cells synthesize basophilic, *keratohyalin granules*, which associate with the tonofibrils. Cells also accumulate *lamellar bodies*, which contain a lipid material that is secreted and serves as a sealant and penetration barrier between cells. Cells also begin to lose other organelles.

- *Stratum lucidum*. A clear layer of non-nucleated, flattened cells that is only visible as a distinct layer in thick skin. In this region, the proteins contained in the keratohyalin granules mediate the aggregation of bundles of keratin filaments (tonofilaments). This process occurs whether or not a distinct stratum lucidum is visible.
- *Stratum corneum*. Variably thick layer of extremely flattened, cornified scales containing aggregated tonofibrils surrounded by a thickened plasma membrane. These cell remnants are sloughed off without damage to the underlying, living epidermal cells.

➢ Other cell types
- Melanocytes
  - Present in stratum germinativum and stratum spinosum
  - Rounded cell bodies with numerous "dendrite-like" processes that insinuate themselves between the keratinocytes
  - Synthesize *melanin*, a dark brown pigment that is packaged into *melanosomes* and injected into keratinocytes
  - Melanin caps the nucleus, reducing damage from solar radiation.
- Langerhans cells. Macrophages that function in immunological skin reactions
- Merkel's cells. Touch receptors.

➢ Epidermal-dermal junction
- Scalloped margin at the interface of the epidermis and dermis, formed by interdigitations of:
  - *Epidermal pegs*. Downward projections of the epidermis
  - *Dermal papillae*. Upward, finger-like protrusion of connective tissue from the dermis
- This junction strengthens the attachment of the epidermis to the underlying dermis.

DERMIS

➢ Composition
- *Papillary layer*
  - Located immediately beneath the basement membrane of the epidermis, forming the dermal papillae
  - Thin layer composed of loose connective tissue
  - Contains small blood vessels, nerves, lymphatics, and the sensory receptors, Meissner's corpuscles

- *Reticular layer*
  - Located between the papillary layer and the hypodermis
  - Thick layer composed of dense, irregular connective tissue
  - Contains larger nerves and blood vessels, glands, hair follicles, and the sensory receptors, Pacinian corpuscles and Ruffini end organs
- Vasculature of the dermis
  - *Papillary plexus* located in the dermal papillae
  - *Cutaneous plexus* located in the reticular layer of the dermis
  - *Arteriovenous anastomoses* allow shunting of blood between papillary and cutaneous plexuses for temperature regulation.

HYPODERMIS

- Not technically part of the integument
- Composed of loose connective tissue and adipose tissue, which can accumulate in large fatty deposits
- Provides anchorage for skin to the underlying tissues
- May contain the bases of sweat glands and hair follicles
- Many sensory receptors, especially Pacinian corpuscles, are present.

STRUCTURES ASSOCIATED WITH THE SKIN

- Glands
  - *Sweat glands*
    - Simple, coiled tubular glands
    - Contain *myoepithelial cells,* which are specialized cells that contract to aid in the expulsion of the sweat
    - Types of sweat glands
      - *Merocrine* or *eccrine*. Located in all regions of the body except the axillary and anal regions; produce a watery secretion that empties onto the surface of the epidermis
      - *Apocrine*. Restricted to the axillary, areolar, and anal regions; much larger than eccrine sweat glands with a broader lumen. Produce a viscous secretion that empties into the hair follicle. Do not secrete by the apocrine mode.
  - *Sebaceous glands*
    - Simple, branched acinar glands
    - Usually secrete into a hair follicle

- ◆ Produce sebum, an oily secretory product, released by the holocrine mode of secretion
- ◆ Absent from thick skin

➢ *Hair follicles*

- Invaginations of the epidermis
- Consist of a *bulb* at the base of the follicle that is located in the hypodermis or in the deep layers of the dermis. *Internal* and *external sheaths* surround the growing *hair shaft* as it passes though the dermis and epidermis.
- An *arrector pili* muscle attaches a hair follicle to the papillary layer of the dermis. Contraction provides elevation of the hair, forming "goose-bumps."

➢ *Nails*

- Keratinized epithelial cells on the dorsal surface of the fingers and toes
- Consist of a *nail plate* that corresponds to the stratum corneum of the epidermis. This plate rests on the *nail bed*, consisting of cells corresponding to the stratum spinosum and stratum germinativum.

➢ Sensory structures

- Nonencapsulated. *Free nerve endings* in the epidermis, responsive to touch, pressure, heat, cold, and pain
- Encapsulated pressure receptors
  - ◆ *Meissner's corpuscle*
    - ■ Located at the apex of a dermal papilla
    - ■ Consists of a coil of *endoneurial cells* around a nerve terminal
    - ■ Responds to light touch
  - ◆ *Pacinian corpuscle*
    - ■ Located in the dermis and hypodermis
    - ■ Consists of concentric layers of endoneurial cells around a nerve terminal
    - ■ Responds to deep pressure
  - ◆ *Ruffini ending*
    - ■ Located in the dermis
    - ■ Consists of groups of nerve terminals surrounded by a thin connective tissue capsule
    - ■ Responds to touch and pressure

Structures Identified in This Section

- Thick skin
- Thin skin
- Epidermis
  - Epidermal pegs
  - Keratin filaments
  - Keratinocyte
  - Stratum corneum
  - Stratum germinativum
    - Melanocyte
    - Melanin granules (melanosome)
  - Stratum granulosum
  - Stratum lucidum
  - Stratum spinosum
- Dermis
  - Collagen bundles
  - Dermal papillae
    - Sensory papilla
      - Meissner's corpuscle
    - Vascular papilla
  - Elastic fibers
  - Papillary layer
  - Reticular layer
- Hypodermis
  - Adipose connective tissue
  - Hair follicle
    - Bulb
    - Papilla
    - Follicle sheath
    - Hair shaft
  - Pacinian corpuscle
    - Perineurial cells
    - Axon
  - Sebaceous gland
  - Sweat gland
    - Duct
    - Secretory portion
      - Myoepithelial cell
      - Secretory granules

CHAPTER

12

DIGESTIVE SYSTEM

ORAL CAVITY

COMPONENTS

- *Vestibule*. Bounded anteriorly and laterally by the lips and cheeks; bounded medially by teeth and gingiva
- *Oral cavity proper*. Bounded anteriorly and laterally by the lingual surfaces of the teeth and gingiva, superiorly by the hard and soft palate, inferiorly by the tongue and floor of the mouth, and posteriorly by the pillars of the fauces leading to the pharynx

ORAL MUCOSA

- Oral mucosa, the mucous membrane lining the oral cavity, is continuous with external skin and with the mucous membrane of the pharynx.
- Composition
  - Epithelium. Stratified squamous keratinized or nonkeratinized depending on location
  - Lamina propria
  - Muscularis mucosae is not present.

Digital Histology: An Interactive CD Atlas with Review Text, by Alice S. Pakurar and John W. Bigbee
ISBN 0-471-64982-1

- Although not part of the oral mucosa, a submucosa of dense connective tissue, containing the minor salivary glands, underlies much of the oral mucosa.

➢ Regional variations
- *Masticatory mucosa*
  - ◆ Located where mucosa is exposed to forces of mastication, such as gingiva and hard palate
  - ◆ Composition
    - ■ Stratified squamous epithelium, keratinized
    - ■ Underlying submucosa is lacking in some locations.
- *Lining mucosa*
  - ◆ Located where mucosa is not exposed to forces of mastication, such as lining of lips and cheeks, soft palate, alveolar mucosa, undersurface of tongue, and floor of mouth
  - ◆ Epithelium. Stratified squamous epithelium, nonkeratinized (moist)
- *Specialized mucosa*
  - ◆ Named "specialized" due to the presence of taste buds
  - ◆ Located on the dorsum of the tongue, where it forms papillae
  - ◆ Epithelium
    - ■ Stratified squamous keratinized, modified to form filiform papillae that facilitate the movement of food posteriorly
    - ■ Stratified squamous moist, covering fungiform and circumvallate papillae

TONGUE

➢ The subdivisions of the *tongue* are based on embryologic origins: anterior two-thirds (body) and posterior one-third (root) are separated by the sulcus terminalis.

➢ Composition
- Mucosa. Dorsum of the tongue is lined by a specialized oral mucosa, modified to form papillae. (See "Specialized mucosa" above.) The ventral surface of the tongue is lined by a lining mucosa.
- The submucosa possesses minor salivary glands that are mucus-secreting except for those associated with the circumvallate papillae, which are serous-secreting.

- *Papillae.* Each consists of a connective tissue core covered by a stratified squamous epithelium.
  - *Filiform*
    - Most numerous; cover body of tongue
    - Cone-shaped protrusions angled so that they aid in movement of food toward the pharynx
  - *Fungiform*
    - Less numerous than filiform but also located on anterior two-thirds of tongue
    - Mushroom shaped, possess taste buds on superior surface
  - *Circumvallate*
    - Eight to twelve papillae located just anterior to the sulcus terminalis
    - Mushroom shaped and surrounded by a narrow moat; lateral wall of papilla possesses taste buds
    - Serous glands of von Ebner open into the base of the moat and flush the moat for reception of new tastes.
  - Foliate. Parallel folds on the posterolateral surface of the tongue; not well developed in humans
- *Taste buds* are onion-shaped structures embedded in the surface of the fungiform and circumvallate papillae. Taste buds contain taste-receptor cells that communicate with the surface of the papilla through a *taste pore.* Depolarization of the taste cells leads to the stimulation of gustatory nerve fibers and the discrimination of sweet, salty, bitter, and sour sensations.
- *Intrinsic tongue muscles.* Skeletal muscle bundles are arranged in three separate planes, with connective tissue bands from the lamina propria separating the bundles and firmly anchoring the muscle to the mucous membrane.

Teeth

- Overview of the *teeth*
  - *Anatomic crown.* The portion of the tooth covered by enamel.
  - *Anatomic root.* The portion of the tooth covered by cementum.
  - *Cervix.* Region where enamel abuts cementum
  - *Pulp cavity* is the central core of a tooth and is divided into a pulp chamber in the crown and a root canal in the root. An apical foramen at the tip of the root allows passage of nerves and blood vessels into and out of the pulp cavity.

- *Gingiva*. Oral mucosa encircling the cervical region of the tooth and providing support for the tooth

➢ Components

- *Enamel* is the hardest tissue in the body, consisting of a mineralized tissue that is 96% hydroxyapatite. Enamel covers the anatomic crown of the tooth. During tooth development, enamel deposition by ameloblasts begins on the surface of dentin and progresses away from this dentinoenamel junction. No additional enamel can be formed after the tooth erupts, as the ameloblasts die on exposure to the oral cavity.
- *Dentin*
    - ◆ Comprises the bulk of the tooth, underlying both enamel and cementum; dentin is a connective tissue that is 70% mineralized with hydroxyapatite.
    - ◆ Dentin is formed continuously throughout life by odontoblasts whose cell bodies line the pulp cavity.
    - ◆ Odontoblast processes extend through the dentin in S-shaped dentinal tubules radiating from the odontoblasts toward the dentinoenamel or dentinocemental junctions.
- *Cementum*, a connective tissue mineralized with 50% hydroxyapatite, covers the anatomic root of the tooth. Cementum is formed continuously throughout life by activity of cementoblasts lying on the surface of the root at the interface of the cementum with the periodontal ligament.
- The *pulp cavity* is lined by odontoblasts and filled with loose connective tissue, blood vessels, nerves, and lymphatics.
- The *periodontal ligament*, collagen fiber bundles interconnecting cementum with the surrounding alveolar bone, suspends and supports each tooth in its alveolar socket.

MAJOR SALIVARY GLANDS

➢ Overview

- All major salivary glands are compound, exocrine glands, and all open into the oral cavity.
- Functions
    - ◆ Produce saliva to wet, lubricate, and buffer the oral cavity and its contents
    - ◆ Produce amylase for the initial digestion of carbohydrates
    - ◆ Produce lysozyme to control bacteria in the oral cavity

- Major cell types
  - *Serous cells*
    - Synthesize, store, and release a thin, protein-rich secretion containing digestive enzymes, primarily amylase
    - Are pyramidal in shape and possess all organelles necessary for protein production and secretion (e.g., basal RER, Golgi, and apical secretory granules)
    - Are arranged into either
      - *Acini* (singular, *acinus*) or *alveoli* (singular, *alveolus*). Flask-shaped sacs with tiny lumens
      - *Serous demilunes*. Half moon–shaped caps positioned over the ends of mucous tubules
  - *Mucous cells*
    - Synthesize, store, and release mucus, a viscous, thick, glycoprotein secretion that protects and lubricates epithelia
    - Have flattened nuclei that are located at the bases of the cells along with the RER. Abundant mucigen droplets are located in the apex of each cell, giving it a frothy, vacuolated appearance.
    - Are organized in test tube–shaped tubules with relatively wide lumens
  - *Myoepithelial cells* are stellate-shaped epithelial cells with contractile functions that lie between the secretory or duct cells and the basement membrane. These cells contract to aid in movement of the secretory product.
- Duct system conducts secretions to oral cavity.
  - Ducts are more numerous with serous acini than with mucous tubules because the tubules can act as their own ducts.
    - *Intralobular ducts*
      - *Intercalated ducts* exit from secretory acini and are smaller in diameter than the acini they drain. These ducts are lined by simple cuboidal epithelia.
      - *Striated ducts* are larger in diameter than the secretory units they drain. They are lined by simple columnar epithelia. Numerous mitochondria and infoldings of the plasma membrane in the basal region of the cells give the duct a striated periphery. Striated ducts alter the content and concentration of the saliva.
    - *Interlobular ducts* are located in the connective septa between lobules and are lined with simple columnar to stratified columnar epithelia.

- ◆ The *main excretory duct(s)* is lined by a stratified epithelium that becomes stratified squamous moist just prior to its junction with the epithelium of the oral cavity.

➢ Major salivary glands

- *Parotid glands*
  - ◆ Compound acinar glands producing only serous products; their secretions account for 25% of the saliva
  - ◆ Possess the most highly developed duct system of the major salivary glands
- *Submandibular glands*
  - ◆ Compound tubulo-acinar glands producing both serous and mucous products, although serous acini predominate. Their secretions account for 70% of the saliva.
  - ◆ Serous cells are present as both acini and serous demilunes.
- *Sublingual glands* secrete approximately 5% of the saliva. These are compound tubulo-acinar glands, producing both mucous and serous products, although mucous tubules predominate.

STRUCTURES IDENTIFIED IN THIS SECTION

Oral Cavity
- Lip
  - Connective tissue
  - Connective tissue papillae
  - Hair
  - Labial glands
  - Labial glands, ducts
  - Mucosa, lining
  - Orbicularis oris
  - Sebaceous glands
  - Skeletal muscle
  - Skin
  - Epithelium, stratified squamous moist
  - Epithelium, stratified squamous keratinized
  - Vermilion zone
- Tooth
  - Cementum
  - Dentin
  - Dentinal tubules
  - Dentino-cementum junction
  - Dentino-enamel junction
  - Enamel
  - Enamel stria
  - Pulp cavity
- Tongue
  - Capillaries
  - Connective tissue core
  - Dorsal surface
  - Epithelium, keratinized
  - Furrow (moat)
  - Papillae, circumvallate
  - Papillae, filiform
  - Papillae, fungiform
  - Serous glands of von Ebner
  - Serous glands of von Ebner, ducts
  - Skeletal muscle
  - Taste buds
  - Taste pores
  - Ventral surface

Major salivary glands
- Connective tissue septa
- Ducts, intercalated
- Ducts, interlobular
- Ducts, striated
- Lobules
- Secretory units

Parotid gland
- Acinar lumens
- Artery, small
- Capillaries
- Ducts, intercalated
- Ducts, interlobular
- Ducts, intralobular
- Ducts, striated
- Epithelium, stratified columnar
- Interlobular connective tissue
- Lobules
- Peripheral nuclei
- Secretory granules
- Serous acini
- Vein, small

Submandibular gland
- Acinar lumens
- Ducts, intercalated
- Ducts, interlobular
- Ducts, intralobular
- Ducts, striated
- Mucous tubules
- Secretory granules
- Serous acini
- Tubular lumens

Sublingual gland
- Ducts, intercalated
- Ducts, interlobular
- Ducts, intralobular
- Lobules
- Mucous cells
- Mucous tubules
- Serous acini
- Serous cells
- Serous demilune

TUBULAR DIGESTIVE SYSTEM

COMPONENTS

- ➢ Pharynx
- ➢ Esophagus

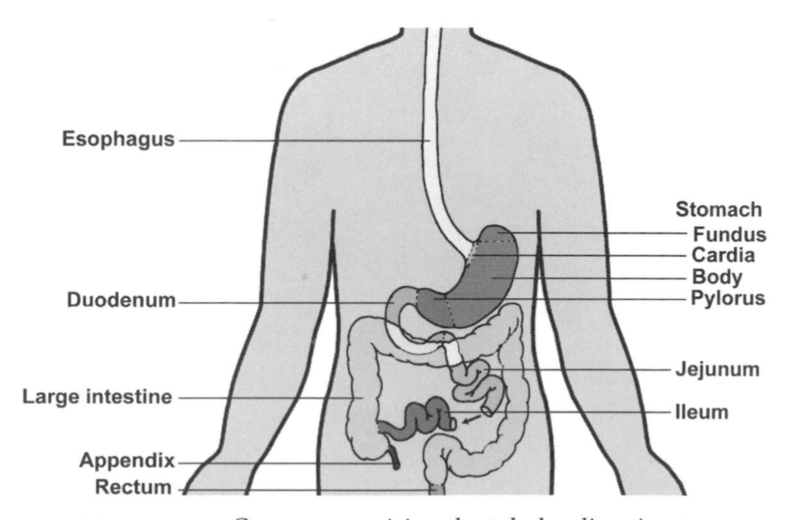

FIGURE 12.1. Organs comprising the tubular digestive tract.

- Stomach
- Small intestine
- Large intestine

BASIC HISTOLOGICAL ORGANIZATION

- Layers
  - *Mucosa* (mucous membrane). Innermost layer facing the lumen
    - Epithelium. Either a stratified squamous moist or a simple columnar epithelium
    - Lamina propria. Loose connective tissue; usually possesses digestive glands
    - Muscularis mucosae of smooth muscle is usually present.
  - *Submucosa*. Denser connective tissue than the lamina propria. The submucosa possesses Meissner's nerve plexus that supplies innervation to the muscularis mucosae and to digestive glands in the mucosa and submucosa. The submucosa possesses glands in the esophagus and duodenum.
  - *Muscularis externa* of smooth muscle is usually arranged into inner circular and outer longitudinal layers. Auerbach's nerve plexus is located between the two muscle layers and provides innervation to this smooth muscle.
  - *Serosa* (serous membrane) is present if the organ protrudes into the peritoneal cavity, or an *adventitia* (only the connective tissue portion of the serosa) is present if the organ is retroperitoneal.

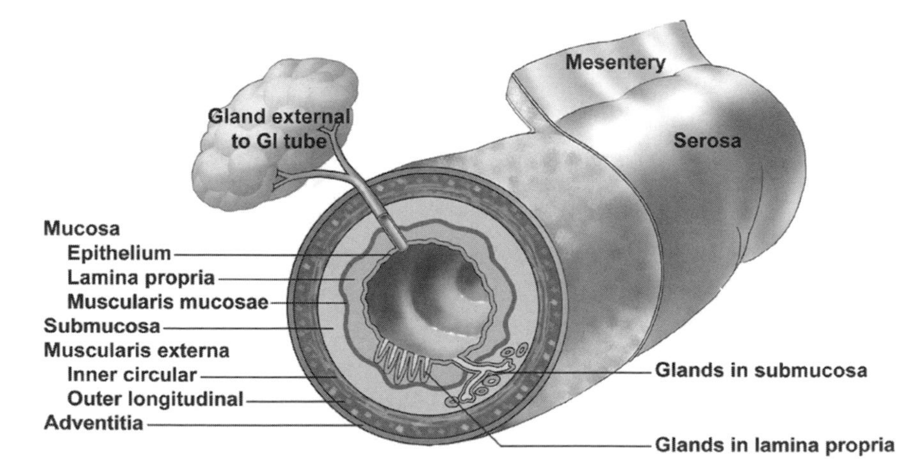

FIGURE 12.2. Overview of the layers and components of the tubular digestive tract.

➢ Glands

- Exocrine glands, aiding in digestion and/or lubrication, are located in:
  - Epithelium (e.g., goblet cells throughout the intestines)
  - Lamina propria (e.g., gastric glands)
  - Submucosa (e.g., Brunner's glands in the duodenum)
  - Glands located external to the digestive tract that open into the system (e.g., liver and pancreas)
- Endocrine and paracrine cells, belonging to the diffuse neuroendocrine system (DNES), are located throughout the mucosa of the gastrointestinal tract, influencing the secretion of glands and the motility of the gut.

Variations That Distinguish Each Organ from the Basic Organizational Plan

➢ Esophagus

- Epithelium. Stratified squamous nonkeratinized epithelium
- Lamina propria possesses esophageal cardiac glands that are mucus-secreting and are particularly prominent near the junction of the esophagus with the stomach.
- Submucosa has mucus-secreting, esophageal glands.
- Muscularis externa is composed of striated muscle in the upper portion of the esophagus, skeletal, and smooth muscle in the middle portion, and smooth muscle in the lower portion.
- Adventitia. Composed of loose connective tissue.

➢ *Stomach*

- Structures present throughout the stomach

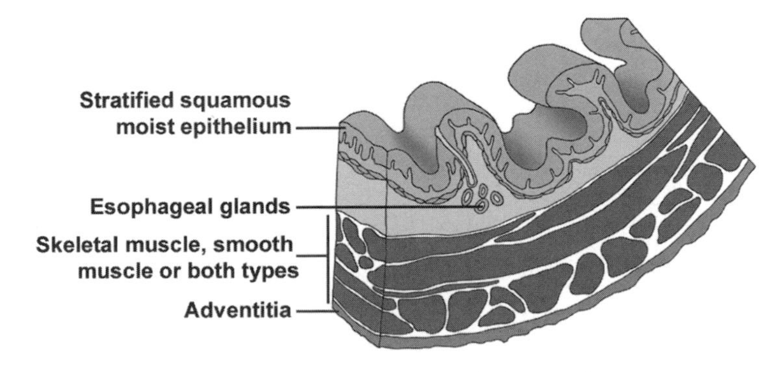

FIGURE 12.3. Cross-section of the esophagus.

- Surface epithelium
  - Simple columnar epithelium facing the lumen is modified so that all cells secrete mucus, forming a *sheet gland* that protects the stomach from its acidic environment.
  - *Gastric pit*. A channel formed by the invagination of the surface epithelium into the underlying lamina propria; connects the sheet gland with the gastric glands. The length of the gastric pit varies with each stomach region.
- *Gastric glands*
  - Simple, branched tubular glands begin at a gastric pit and extend through the lamina propria to the muscularis mucosae.
  - The region of the gland that attaches to the gastric pit is called the neck region; the base region of the gland is located adjacent to the muscularis mucosae.
  - Secretory cells in these glands vary in each region of the stomach.
- Muscularis externa. Subdivisions of this layer frequently interdigitate, making it difficult to distinguish one layer from another.
  - Internal oblique layer
  - Middle circular layer that is modified in the pyloric region to form the pyloric sphincter
  - Outer longitudinal layer is separated from the inner circular layer by Auerbach's plexus, nerve fibers from the autonomic nervous system that supply muscularis externa.

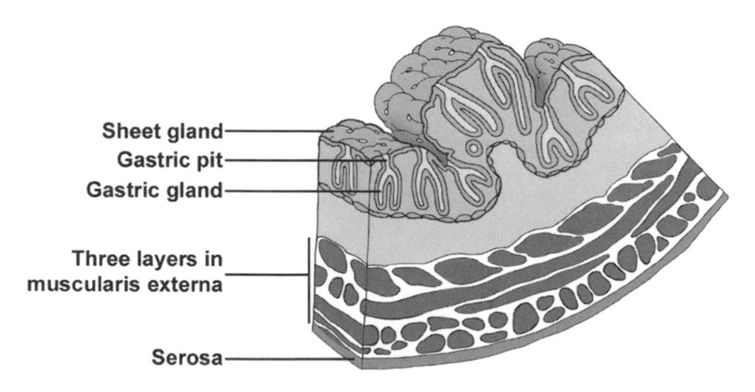

FIGURE 12.4. Cross-section of the stomach.

  - Serosa
  - *Rugae*. Longitudinal folds of the mucosa and submucosa in the undistended stomach allow for expansion.
- Variations specific to the *cardiac region* (narrow region adjacent to the esophagus)
  - Abrupt transition of epithelium from stratified squamous moist of the esophagus to a sheet gland lining the cardiac stomach
  - Length of gastric pits is about equal to the length of cardiac glands.
  - *Cardiac glands* primarily secrete mucus, although other products are also produced. Glands are frequently coiled.
  - Cardiac glands of the stomach extend into the lower esophagus, forming the esophageal cardiac glands.
- Variations specific to the *fundic* and *body regions* (Glands in both regions are called fundic glands.)
  - *Fundic glands* are about twice as long as their gastric pits.
  - Cell types present:
    - *Stem cells* replenish both the surface epithelial cells and cells of the glands. Stem cells are located in the neck region.
    - *Mucous neck cells* are irregular in shape and stain basophilically. They secrete mucus and are located in the neck region.
    - *Parietal cells* are large, spherical, eosinophilic cells that secrete hydrogen and chloride ions and gastric intrinsic factor. They possess numerous mitochondria. An umbrella-shaped canaliculus indents the luminal surface, increasing surface area. Although present throughout the gland, parietal cells are more numerous in the upper regions.
    - *Chief* or *zymogen cells*, typical protein-producing cells, predominate in the bases; stain blue with hematoxylin and secrete pepsinogen.
    - *Enteroendocrine cells* (part of the diffuse neuroendocrine system, DNES) are located on the basement membrane and do not usually reach the lumen of the gland. This population of cells secretes a variety of hormones with endocrine and paracrine influences on digestive activity. Secretory granules cluster toward the basement membrane for their subsequent release into the lamina propria. Most common at the bases of the glands.
- Variations specific to the *pyloric region*

    - Pits are longer in pylorus than in the cardiac region.
    - *Pyloric glands*, not as coiled as in the cardiac region; primarily secrete mucus.
    - Enteroendocrine cells are also present here.
    - Circular layer of muscularis externa is greatly thickened to form the *pyloric sphincter*.

- *Small intestine*
  - Subdivided into duodenum, jejunum, and ileum
  - Common features of the small intestine
    - Structures that increase the surface area of the small intestine
      - *Microvilli*. Increase surface area of absorptive cells and, collectively, form a brush or striated border
      - *Villi*. Finger-like protrusions of the lamina propria and overlying epithelium into the lumen
        - Villi assume different shapes in each of the three intestinal subdivisions.
        - A lacteal (blind-ending lymphatic capillary) is located in the center of each villus to absorb digested fat.
        - Individual smooth muscle cells lie parallel to the long axis of each villus, "milking" the lacteal contents to the periphery.
      - *Plicae circulares*. Permanent circular folds formed by an upwelling of the submucosa and its overlying mucosa into the lumen. Villi protrude from the plicae.

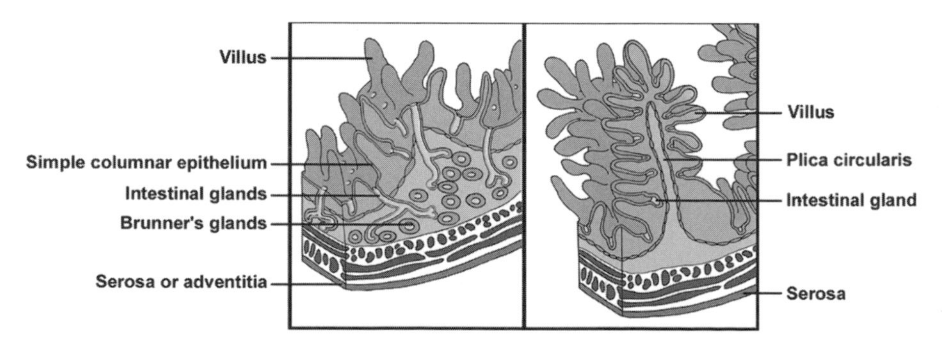

FIGURE 12.5. Longitudinal section through the duodenum (left) and the jejunum/ileum (right). Note the orientation of the layers of muscularis externa when sectioned longitudinally.

- Mucosal epithelium is composed of:
  - *Absorptive cells*, forming a simple columnar epithelium with microvilli, absorb digested food
  - *Goblet cells* (unicellular glands) are interspersed among absorptive cells and secrete mucus. These cells increase in number from duodenum to rectum.
- *Intestinal glands* (*crypts of Lieberkuhn*) are simple tubular glands that begin at the bases of the villi in the mucosa and extend through the lamina propria to the muscularis mucosae. Possess:
  - Absorptive cells
  - Goblet cells
  - *Paneth cells* possess large, eosinophilic granules whose contents digest bacterial-cell walls.
  - *Enteroendocrine cells*
- Muscularis externa of inner circular and outer longitudinal layers with an intervening Auerbach's nerve plexus
- Serosa covers all of small intestine except for the beginning of the duodenum, which is retroperitoneal and possesses an adventitia.

- Variations specific to the intestinal subdivisions
  - *Brunner's glands* in the submucosa are present only in the duodenum. These compound tubular glands open into the bases of the intestinal glands and secrete an alkaline mucus to neutralize the acidity of the stomach contents.
  - *Peyer's patches* are clusters of 10–200 lymphoid nodules located primarily in the lamina propria of the ileum. Each cluster is positioned on the side of the intestine away from the mesentery and forms a bulge that may protrude into the lumen as well as into the submucosa.

➢ *Large intestine (colon)*

- Mucosal epithelium:
  - Absorptive cells form a simple columnar epithelium with microvilli.
  - Goblet cells increase in number toward the rectum and provide lubrication.
  - A reduced number of enteroendocrine cells is present.
- Intestinal glands (crypts of Lieberkuhn) are very straight in the large intestine.
- No villi or plicae circulares are present in the large intestine.

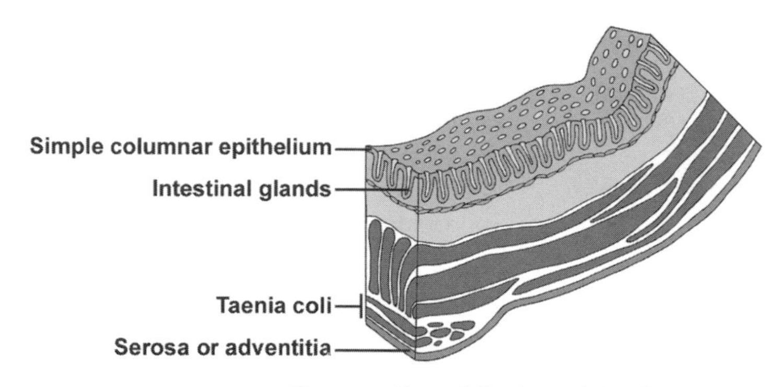

FIGURE 12.6. Cross-section of the large intestine.

- Muscularis externa
  - Inner circular layer is intact.
  - Outer longitudinal layer is segregated into three longitudinal bands, the *taeniae coli,* that are placed equidistantly around the tube. The contraction of the taenia produces permanent sacculations in the large intestine, termed *haustrae.*
- Either an adventitia or a serosa is present, depending on the particular portion of the large intestine.
- The *appendix* resembles the large intestine except that the outer longitudinal smooth muscle layer is intact. Additionally, abundant lymphoid tissue is present in the lamina propria to protect against invading microorganisms.
- *Rectum* is a 12-cm-long tube continuing from the sigmoid colon. The mucosa of the rectum is similar to that of the majority of the large intestine. The rectum narrows abruptly to become the anal canal.
- *Anal canal.* The terminal portion of the intestinal tract is about 4 cm long.
  - The intestinal glands disappear and the epithelium undergoes an abrupt transition from simple columnar to stratified squamous with sebaceous and apocrine sweat glands.
  - The inner circular portion of the muscularis externa expands to form the internal anal sphincter. The external anal sphincter is composed of skeletal muscle.

Structures Identified in This Section

Overview of tubular digestive tract
- Adventitia
- Epithelium
- Esophagus
- Glands
- Lamina propria
- Large intestine
- Mucosa
- Muscularis externa
- Muscularis mucosae
- Plicae circulares
- Serosa
- Small intestine
- Stomach
- Submucosa

Esophagus
- Adventitia
- Capillaries
- Cardiac glands
- Cardiac stomach
- Epithelium, stratified squamous moist
- Esophageal glands
- Gastric pits
- Gastric pits, openings
- Gastroesophageal junction
- Lamina propria
- Mucosa
- Muscularis externa
- Muscularis mucosae
- Skeletal muscle (ls and xs)
- Smooth muscle
- Squamous cells
- Submucosa

Stomach
- Auerbach's plexus
- Brunner's glands
- Cardiac glands
- Chief cells
- Collagen fibers
- Enteroendocrine cells
- Fundic glands
- Gastric glands
- Gastric glands, bodies
- Gastric glands, branching
- Gastric glands, necks
- Gastric pits
- Gastroduodenal junction
- Intestinal glands
- Lamina propria
- Lymphoid nodule
- Mast cells
- Meissner's plexus
- Mucosa
- Mucous neck cells
- Mucigen
- Muscularis externa
- Muscularis externa, middle circular layer
- Muscularis externa, outer longitudinal layer
- Muscularis mucosae
- Neuron cell body
- Parietal cells
- Peripheral nerves
- Plasma cells
- Pyloric glands
- Pyloric sphincter
- Secretory granules
- Serosa
- Sheet gland
- Sheet gland, stem of cell
- Stomach, cardiac
- Stomach, fundic and body
- Stomach, pyloric
- Submucosa
- Villi

Small intestine
- Absorptive cells (enterocytes)
- Brunner's glands
- Brunner's glands, ducts
- Duodenum
- Enteroendocrine cells

Epithelium, simple columnar with microvilli
Epithelium, simple squamous
Germinal center
Goblet cells
Intestinal glands
Jejunum/ileum
Junctional complex
Lacteal
Lamina propria
Lumen
Lymphoid nodule
Mesentery
Microvilli
Mucosa
Muscularis externa
Muscularis mucosae
Paneth cells
Peyer's patch
Plasma cells
Plicae circulares
Serosa
Submucosa
Villi

Large intestine
- Anal canal
- Anal sphincter, external
- Anal sphincter, internal
- Anus
- Apocrine sweat glands
- Diffuse lymphoid tissue
- Hair follicle
- Intestinal glands
- Lymphatic nodules
- Mucosa
- Muscularis externa
- Muscularis externa, inner circular layer
- Muscularis externa, outer longitudinal layer
- Recto-anal junction
- Rectum
- Serosa
- Submucosa
- Taeniae coli

MAJOR DIGESTIVE GLANDS

PANCREAS

OVERVIEW

➢ Located in the abdomen in the curve of the duodenum and divided into a head, body, and tail

➢ Is both an exocrine and an endocrine gland

- The exocrine portion produces an alkaline secretion containing digestive enzymes that empties into the duodenum.
- The endocrine portion secretes insulin, glucagon, and somatostatin that regulate blood glucose levels.

MICROSCOPIC ANATOMY

➢ *Exocrine pancreas*

- Compound acinar gland; the acinar cells secrete numerous digestive enzymes that breakdown proteins, carbohydrates, lipids, and nucleic acids.

- Cells show polarity with basal RER and apical secretory granules.
- Duct system
  - Ducts begin as *centroacinar cells* located within the acini.
  - Intercalated ducts are lined with simple cuboidal epithelium. Centroacinar cells and cells of the intercalated ducts secrete bicarbonates to neutralize the acidity of the stomach contents (chyme) entering the duodenum.
  - Striated ducts are not present.
  - Interlobular ducts lead into one or more excretory ducts that empty into the duodenum.
- Resembles the parotid gland except the pancreas has centroacinar cells and fewer ducts.
- Secretion is regulated by cholecystokinin and secretin from enteroendocrine cells in the small intestine.

➢ *Endocrine pancreas (islets of Langerhans)*

- Small clusters of cells, richly supplied by fenestrated capillaries, are scattered throughout the exocrine pancreas; these clusters show no orderly arrangement of secretory cells within the cluster.
- Predominate cell types and secretions
  - *A cell* (alpha cell). Secretes glucagon, which elevates glucose levels in the blood
  - *B cell* (beta cell). Secretes insulin, which lowers blood glucose levels; predominant cell type
  - *D cell* (delta cell). Secretes somatostatin, which modulates release of the other two major hormones
- Individual cell types cannot be distinguished with hematoxylin and eosin.

LIVER

OVERVIEW

➢ Located in right, upper quadrant of abdominal cavity under the diaphragm

➢ Both an exocrine and an endocrine gland

- Exocrine secretion (bile) is stored in the gall bladder and released into the duodenum. This secretory product contains bile acids that aid in the emulsification of lipids, bilirubin (the breakdown product of hemoglobin), phospholipids, and cholesterol.

- Endocrine function is the synthesis of plasma proteins, including albumin, clotting factors, and lipoproteins that are released into the liver sinusoids.

➢ Additional functions include the metabolization of digested food, storage of glucose as glycogen, and detoxification.

CYTOARCHITECTURE OF THE CLASSIC LIVER LOBULE

➢ The classic *liver lobule* resembles a column similar to a stack of covered-wagon wheels.

➢ Spokes of the wheels are cords or plates of *hepatocytes* radiating out from a central axis.

➢ Spaces between the spokes are occupied by *sinusoids* (discontinuous sinusoidal capillaries).

➢ Central axis of the lobule is a *central vein* into which sinusoids drain (i.e., blood is flowing from the periphery of the lobule to the center). The central vein runs parallel to the long axis of the lobule.

➢ The perimeter of the lobule (the wheel rim) is difficult to distinguish in the human. The perimeter is denoted by the position of three to six *portal canals* (*hepatic portal triads*) situated at intervals around the lobule.

- Portal canals run parallel to the long axis of the lobule.
- Portal canals contain branches of the
  - ◆ Hepatic portal vein. Lined with simple squamous epithelium; has the largest diameter of the three structures
  - ◆ Hepatic artery. Lined with simple squamous epithelium and two-three layers of smooth muscle
  - ◆ Bile duct. Lined with simple cuboidal epithelium; multiple branches may be present

BLOOD SUPPLY AND DRAINAGE OF THE LIVER

➢ Blood supply is from two sources

- *Hepatic portal vein*
  - ◆ Supplies about 75% of the blood
  - ◆ Carries blood drained directly from the gastrointestinal tract, which, therefore, is deoxygenated and high in absorbed nutrients.
- *Hepatic artery*. Supplies oxygenated blood

➢ Branches from both vascular sources continue into smaller branches located in the portal canals. Portal canal branches supply the hepatic sinusoids that drain into a central vein. Multiple central veins anastomose to eventually form the three hepatic veins that empty into the inferior vena cava.

Functional Microanatomy

➢ Sinusoids

- A variation of discontinuous capillaries, in that gaps exist between endothelial cells and the fenestrations lack diaphragms
- The basal lamina is lacking beneath the fenestrations.
- Fenestrations open into a subsinusoidal space, the *space of Disse*, separating the sinusoids from the hepatocytes beneath the space.
- *Kupffer cells*, liver macrophages, span the sinusoids, filtering debris from the blood.

➢ *Hepatocytes (liver cells)*

- Arranged as walls one to two cells thick that radiate out from the central vein like the spokes of a wheel
- Histology
  - ◆ Cells are polyhedral in shape.
  - ◆ Cells possess one or two nuclei.
  - ◆ Cells contain abundant smooth and rough endoplasmic reticulum, Golgi apparatus, mitochondria, and lysosomes. They contain large accumulations of electron-dense glycogen granules that stain strongly with PAS. Numerous peroxisomes, along with smooth endoplasmic reticulum, carry out detoxification.
  - ◆ At intervals between adjacent cells, the plasma membranes bulge inward to form a bile canaliculus, the beginning of the bile transport system.
  - ◆ Microvilli project into the space of Disse, increasing surface area of the cells.

Flow of Bile from Liver

➢ Bile is produced by hepatocytes in the liver and released into bile canaliculi located between two adjacent hepatocytes.

➢ Bile canaliculi form a meshwork configuration that drain into bile ducts lying in portal canals. These bile ducts anastomose to form the hepatic duct.

➢ The hepatic duct exits from the liver. The bile it contains can either:

- Travel directly via the common bile duct, a direct continuation of the hepatic duct, to the duodenum

- Be transported via the cystic duct to the gall bladder where it is stored until needed

GALL BLADDER

OVERVIEW

- Stores and concentrates bile produced in the liver
- Connects, via the *cystic duct*, with the *hepatic duct* from the liver to form the *common bile duct* that empties into the duodenum

MICROANATOMY

- Mucosa
  - Composed of
    - Simple columnar epithelium with short microvilli. Accumulations of mitochondria, particularly in the apices of the cells, are prominent.
    - Lamina propria
    - Muscularis mucosae is not present.
  - Is thrown into complex, irregular folds that are particularly evident when the gall bladder is empty.
- Submucosa
- Smooth muscle is arranged in an irregular network surrounding the gall bladder.
- A serosa covers most of the gall bladder; an adventitia surrounds the portion that is attached to the liver.

STRUCTURES IDENTIFIED IN THIS SECTION

Pancreas
- Acinar cell
- Acini
- Capillaries
- Centroacinar cells
- Ducts, interlobular
- Ducts, intralobular
- Endocrine pancreas
- Exocrine pancreas
- Interlobular connective tissue
- Islets of Langerhans
- Lobules
- RER
- Secretory granules

Liver
- Bile canaliculus
- Bile ducts
- Capsule
- Cell membranes
- Central vein
- Connective tissue

Endothelial cells
Fenestrations
Glycogen
Hepatic arteries
Hepatic portal veins
Hepatic sinusoids
Hepatocyte nucleus
Hepatocytes
Kupffer cells
Liver lobule
Liver sinusoids
Lymphatic vessels
Microvilli
Mitochondria
Nucleus
Portal canals
RBCs
RER
Sinusoids
Space of Disse
Tight junctions

Gall bladder
  Apical granules
  Arteriole
  Connective tissue
  Epithelium, simple columnar with microvilli
  Lamina propria
  Muscle, smooth

CHAPTER

13

RESPIRATORY SYSTEM

OVERVIEW

COMPONENTS OF THE RESPIRATORY SYSTEM

- In relationship to lungs (listed in order from exterior to interior, i.e., the path of inspired air)

Extrapulmonary	Intrapulmonary
1. Nasal cavity	1. Secondary bronchi
2. Pharynx	2. Bronchioles
3. Larynx	3. Terminal bronchioles
4. Trachea	4. Respiratory bronchioles
5. Primary bronchi	5. Alveolar ducts
	6. Alveoli

- According to function (listed in order from exterior to interior)

Conducting Portion (Transports air from exterior)	Respiratory Portion (Involved with gas exchange)
1. Nasal cavity	1. Respiratory bronchioles
2. Pharynx	2. Alveolar ducts
3. Larynx	3. Alveoli
4. Trachea	
5. Primary bronchi	
6. Secondary bronchi	
7. Bronchioles	
8. Terminal bronchioles	

Digital Histology: An Interactive CD Atlas with Review Text, by Alice S. Pakurar and John W. Bigbee
ISBN 0-471-64982-1

STRUCTURE OF "TYPICAL" RESPIRATORY PASSAGEWAYS

- Conducting portion (nasal cavities through secondary bronchi)
  - Mucosa (mucous membrane). Faces the lumen
    - *Respiratory epithelium. Pseudostratified* with *cilia* and *goblet cells.*
    - *Lamina propria* of loose connective tissue with blood vessels and nerves
    - Deepest layer of mucosa may consist of:
      - An *elastic lamina* or
      - A *muscularis mucosae* or
      - This layer may be absent.
  - *Submucosa.* Dense irregular connective tissue with mucous and serous (mixed) glands
  - *Cartilage* or *bone*
  - *Adventitia.* Loose connective tissue forming the outer layer of the passageway
- Structural transitions in walls and layers of the passageways from extrapulmonary passageways to alveoli
  - Transitions
    - Layers become thinner as passageways decrease in diameter.
    - Epithelium decreases in height from pseudostratified to simple columnar to simple cuboidal to simple squamous.
    - Goblet cells and mixed glands stop relatively abruptly at the junction of a secondary bronchus with a bronchiole.

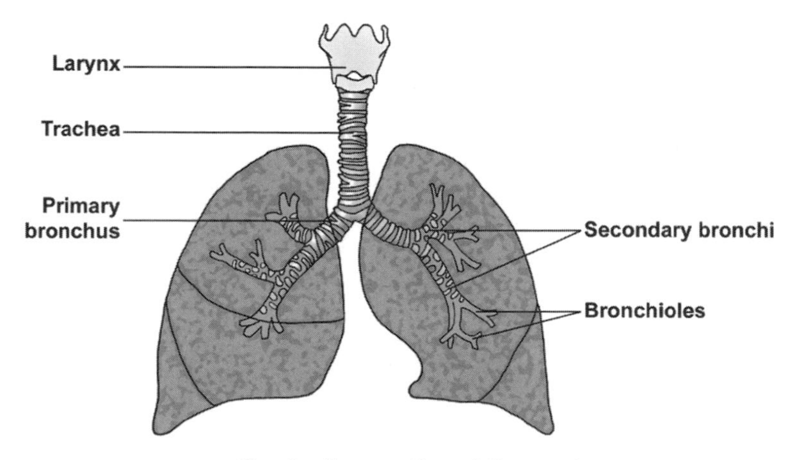

FIGURE 13.1. Conducting portion of the respiratory system.

- Cartilage decreases in size, breaks up into plates, and stops relatively abruptly at the junction of a secondary bronchus with a bronchiole.
- Cilia are gradually eliminated.

- Results in the formation of the wall of an *alveolus,* where gas exchange occurs
  - Epithelium is simple squamous
  - Connective tissue core with numerous capillaries

Conducting Portion

Nasal Cavities (Extrapulmonary)

➢ The *nasal cavities* can be subdivided into two regions, the olfactory and the non-olfactory regions.

➢ *Non-olfactory region*

- *Vestibules.* The epithelium undergoes a transition from epidermis of skin with hairs to pseudostratified, respiratory epithelium.
- *Nasal fossae*
  - "Typical" mucosa but deepest layer is lacking (i.e., neither a muscularis mucosae nor an elastic lamina is present)
  - Patency maintained by bones or cartilage.

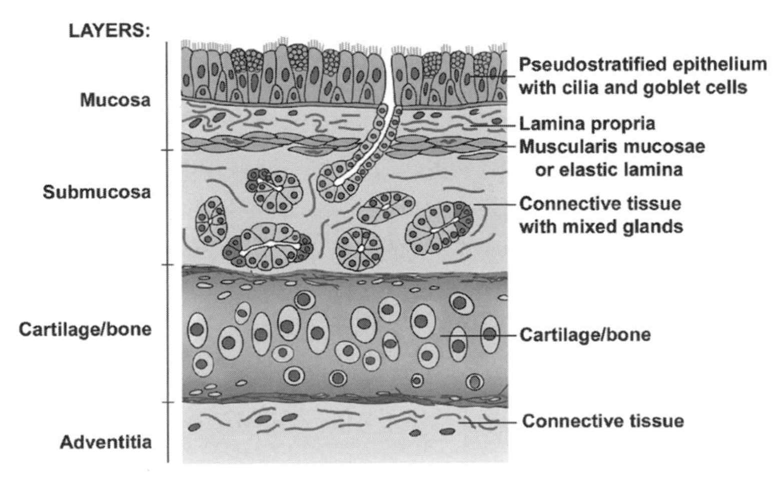

FIGURE 13.2. Layers and components of a "typical" respiratory passageway.

- *Olfactory region*
    - Upside-down, U-shaped area in posterior, superior region of each nasal fossa, extending over superior conchae and about 1 cm down nasal septum
    - Composition of wall
        - Mucosa
            - Epithelium is tall, thick, pseudostratified columnar with nonmotile cilia composed of
                - *Olfactory cells* (neurons). Bipolar neurons that respond to odors. A single dendrite extends to the surface to form a swelling, the *olfactory vesicle*, from which nonmotile cilia extend over the surface. These cilia increase surface area and respond to odors.
                - Support cells span the epithelium and support the olfactory cells.
                - Basal cells are located on the basal lamina and serve as reserve cells for the epithelium.
            - Deepest layer of mucosa is not present, so the lamina propria blends with the submucosa. This connective tissue layer contains *Bowman's glands*, serous glands whose watery secretions flush odorants from the epithelial surface.
        - Patency maintained by bone.

LARYNX (EXTRAPULMONARY)

- Composition of the wall of the *larynx*
    - Mucosa
        - Epithelium
            - Pseudostratified with cilia and goblet cells in most areas
            - Stratified squamous moist over true vocal folds and much of epiglottis because of friction incurred in these areas
        - No muscularis mucosae or elastic lamina, so lamina propria is continuous with submucosa.
    - Submucosa with mixed glands (except in true focal fold)
    - Cartilages maintaining patency are numerous, uniquely shaped and are either hyaline or elastic. The larger cartilages are the epiglottis, thyroid, and cricoid.
    - An adventitia is present.

- Vocal apparatus. Modification in the larynx composed of two pairs of horizontally positioned mucosal folds located on the lateral walls of the larynx
  - *False vocal folds.* More superior in location. Resemble the wall of a "typical" respiratory passageway except the deepest layer of the mucosa is absent (no muscularis mucosae nor elastic lamina).
  - The *ventricle*, a space, separates the false from the true vocal folds.
  - *True vocal folds*
    - Are lined by a stratified squamous moist epithelium and its lamina propria
    - A *vocal ligament* of dense regular elastic connective tissue is located at the edge of the fold, keeping the rim of the fold taut.
    - *Vocalis muscle*, skeletal muscle, lies beneath each true vocal fold. This muscle alters the shape of the vocal fold and aids in phonation.

TRACHEA AND PRIMARY BRONCHI (EXTRAPULMONARY)

- The *trachea* and *primary bronchi* are identical in structure and will be considered together.
- Mucosa
  - Epithelium pseudostratified with cilia and goblet cells with a very prominent basement membrane
  - Lamina propria of loose connective tissue
  - Elastic lamina of longitudinally arranged elastic fibers
- Submucosa with mixed glands
- C-shaped cartilage rings maintain patency; trachealis muscle (smooth) interconnects the open ends of the tracheal rings.
- Adventitia is present.

SECONDARY BRONCHI (INTRAPULMONARY)

- The *secondary bronchi*
  - Are the first intrapulmonary structures; a secondary bronchus supplies each of the three lobes of the right lung and the two lobes of the left lung.
  - Are similar to, but diminished in size from, the primary bronchi
- Mucosa
  - Epithelium, pseudostratified with cilia and goblet cells

    - Lamina propria contains numerous, longitudinally arranged elastic fibers.
    - Muscularis mucosae of smooth muscle fibers arranged in crisscrossing bands
- Submucosa with mixed glands
- Patency maintained by *plates* of *hyaline cartilage.*
- Adventitia is present.

BRONCHIOLES (INTRAPULMONARY)

- Walls of *bronchioles* continue to decrease in size. The greatest changes in histology occur in the walls of the bronchioles as glands and cartilage are eliminated.
- Mucosa
    - Epithelium
        - Pseudostratified with cilia and goblet cells in largest bronchioles that decreases to:
        - Simple columnar with cilia in smallest bronchioles (*terminal bronchioles*), but no goblet cells persist.
        - *Clara cells* are present in terminal bronchioles.
            - Tall, dome-shaped, nonciliated cells
            - Possess numerous secretory granules whose contents aid in lowering surface tension of the terminal bronchioles, thus aiding in inspiration
    - Lamina propria contains numerous, longitudinally arranged elastic fibers.
    - Muscularis mucosae. Greatest development of smooth muscle (crisscrossing bands) in relationship to thickness of wall of all respiratory passageways
- Submucosa contains no glands.
- No cartilages or bones support bronchioles; therefore, submucosa and adventitia form a single connective issue layer.

RESPIRATORY PORTION

PRIMARY FUNCTION

- Gas exchange occurs in the alveolus. Therefore, an alveolus must be an integral part of all the passageways of the respiratory part of the respiratory system.

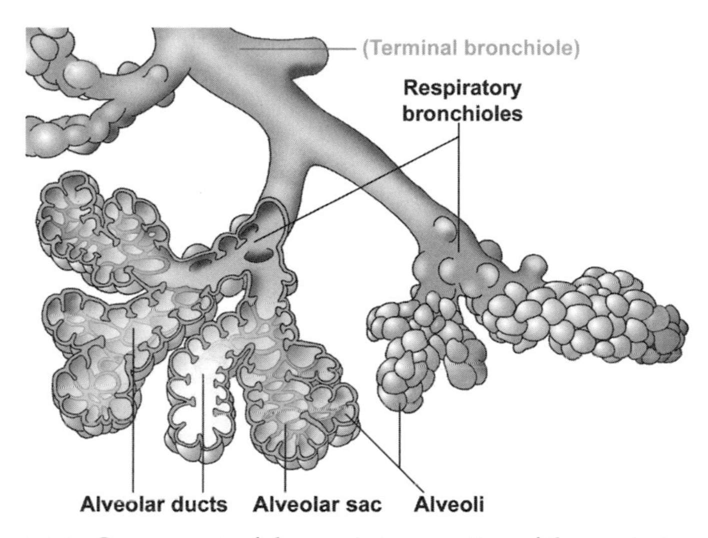

FIGURE 13.3. Components of the respiratory portion of the respiratory system.

Respiratory Bronchioles

- *Respiratory bronchioles* continue to decrease in diameter and in thickness of their walls.
- Mucosa. Simple cuboidal epithelium with a few sparsely scattered cilia and Clara cells
  - Elastic fibers in lamina propria
  - Muscularis mucosae of smooth muscle
- Alveoli bulge from wall (i.e., lumen of alveolus is continuous with lumen of respiratory bronchiole).

Alveolar Ducts

- An *alveolar duct* is formed as the alveoli in a respiratory bronchiole increase in number, thereby decreasing the amount of wall that is present.
  - At the level of the alveolar duct, the "wall" is reduced to a series of rings framing the entrance to an alveolus or a group of alveoli (alveolar sac).
  - When sectioned, these rings resemble knobs to which the alveoli are attached.
- Wall
  - Simple cuboidal epithelium

- Elastic fibers and smooth muscle in "knobs"
- Alveoli bulge from the framework formed by the knobs.

ALVEOLAR SACS

➢ *Alveolar sacs* are two or more alveoli arising from a single ring of knobs.

ALVEOLI

➢ *Alveoli* are thin-walled, hollow polyhedrons forming the bulk of the lungs; where gas exchange occurs.

➢ Alveoli are components of respiratory bronchioles and alveolar ducts, or they may be grouped together to form alveolar sacs.

➢ *Interalveolar septum.* Structure between two adjacent alveoli is composed of:

- The epithelium lining each alveolus.
    - ◆ *Squamous alveolar* or *type I* cells form a simple squamous epithelium lining 95% of the alveolar surface area and forming a portion of the blood-air barrier.

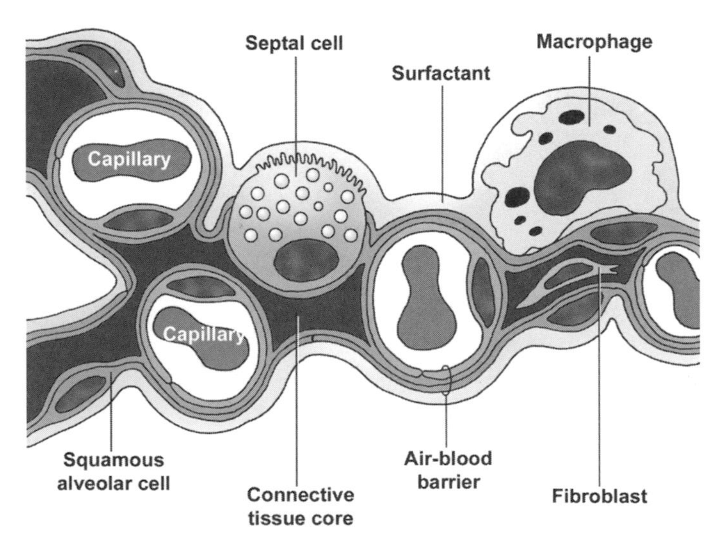

FIGURE 13.4. Components of the interalveolar septum, including the air-blood barrier.

- ◆ *Septal* or *type II* cells
  - ■ Spherical cells with microvilli and abundant, vacuolated cytoplasm; bulge into alveolar space
  - ■ Lamellar bodies in the cytoplasm of septal cells are responsible for the vacuolated appearance of these cells. Lamellar bodies give rise to surfactant, a secretory product consisting of phospholipids, glycosaminoglycans, and proteins.
  - ■ Serve as progenitors for both type I and type II cells
- Connective tissue core contains a vast capillary bed that bulges into the alveolar space, elastic fibers, alveolar macrophages, and other connective tissue components.
- Additional components
  - ◆ Pulmonary *surfactant*
    - ■ Is an extracellular fluid coating alveolar surfaces
    - ■ Lowers alveolar surface tension, aiding in inflation of alveoli during inspiration, and preventing collapse of alveoli during expiration
    - ■ Is composed of a monomolecular, phospholipid surface film that covers an underlying aqueous hypophase
    - ■ Appears during the last weeks of gestation. Absence or insufficiency of surfactant may result in respiratory distress syndrome or hyaline membrane disease in infants born prematurely.
  - ◆ *Alveolar macrophages*
    - ■ Lie free in the alveolar space within the surfactant layer. With congestive heart failure, RBCs pass into alveolar spaces and are phagocytized by these macrophages, which are then called "heart failure" cells.
    - ■ Are located within the connective tissues of all respiratory passageways. Macrophages engulf dust and carbon particles and are called dust cells.
  - ◆ *Alveolar* or *Kohn's pores*. Small openings in the interalveolar septa between neighboring alveoli that aid in equalizing interalveolar pressure. These pores can contribute to the spread of bacteria in the lung.

AIR-BLOOD BARRIER

➢ The *air-blood barrier* separates air from blood. Oxygen and carbon dioxide must cross this barrier during gas exchange.

➢ Composition
  - Squamous alveolar cell with its basement membrane
  - Capillary endothelial cell with its basement membrane

PLEURA

➢ The *pleura* is a serous membrane (serosa) covering the lungs.

➢ Composition
  - Simple squamous epithelium (mesothelium)
  - Underlying connective tissue layer with elastic fibers

➢ Produces a fluid film that lubricates the surface of the lungs and provides surface tension for lung expansion

VASCULAR SUPPLY TO LUNGS

➢ Pulmonary circulation supplies deoxygenated blood for gas exchange
  - The *pulmonary artery* and its branches travel adjacent to the bronchial passageways, supplying deoxygenated blood to the pulmonary capillaries. Pulmonary arteries are comparable in diameter to their neighboring respiratory passageways.
  - *Alveolar (pulmonary) capillaries* lie in the interalveolar septa, forming part of the air-blood barrier. These abundant capillaries anastomose to form pulmonary veins.

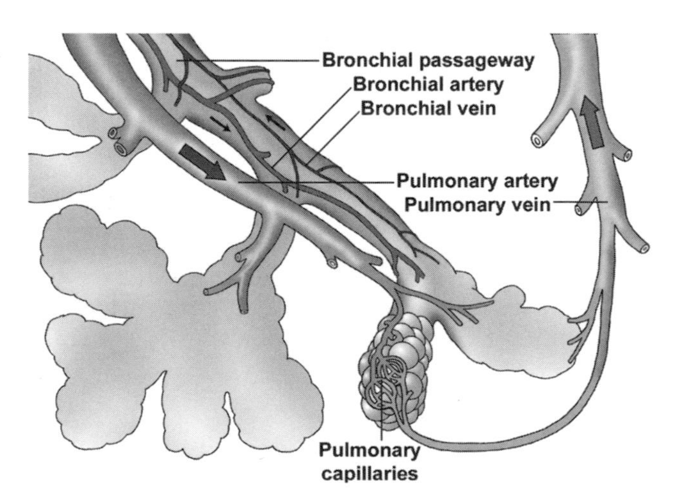

FIGURE 13.5. Vascular supply to the lungs. Arrows indicate direction of blood flow.

- *Pulmonary veins* carry oxygenated blood, and travel alone in the parenchyma away from respiratory passageways. After leaving a lung lobule, pulmonary veins, the bronchial passageways, and pulmonary arteries collect near the hilum of the lung.

➢ *Bronchial circulation* supplies oxygenated blood and nutrients to the tissue layers forming the walls of the bronchial passageways. These vessels, therefore, lie within and are much smaller than the wall of the passageways they supply.

STRUCTURES IDENTIFIED IN THIS SECTION

Overview
- Adventitia
- Air-blood barrier
- Alveolar spaces
- Basement membrane
- Capillaries
- Cartilage
- Cilia
- Connective tissue core
- Deepest layer
- Elastic lamina
- Endothelium
- Epithelium, respiratory
- Epithelium, simple squamous
- Goblet cells
- Interalveolar septa
- Lamina propria
- Mucosa
- Submucosa
- Surfactant

Nasal cavity
- Bone
- Bowman's glands
- Cartilage
- Cartilage/bone
- Conchae
- Connective tissue
- Epithelium, olfactory
- Epithelium, pseudostratified columnar
- Nasal cavities
- Nasal septum
- Olfactory mucosa
- Respiratory mucosa
- Skin

Larynx
- Cartilages
- Epithelium
- Epithelium, respiratory
- Epithelium, stratified squamous
- False vocal fold
- Trachea
- True vocal folds
- Ventricle
- Vocal ligament
- Vocalis muscle

Trachea and primary bronchus
- Basement membrane
- Cartilage
- Cartilage ring
- Cartilage ring opening
- Elastic lamina
- Epithelium
- Epithelium, respiratory
- Esophagus
- Lamina propria
- Mixed glands
- Mucosa
- Perichondrium
- Submucosa
- Trachea
- Trachealis muscle

Secondary bronchus
- Alveoli
- Bronchial blood vessels

- Bronchiole
- Cartilage plates
- Elastic fibers
- Epithelium, respiratory
- Lamina propria
- Macrophages
- Mixed glands
- Muscularis mucosae
- Pulmonary vessels
- Secondary bronchus

Bronchioles
- Alveolar ducts
- Alveoli
- Basal bodies
- Bronchial blood vessels
- Bronchioles
- Cilia
- Clara cells
- Elastic fibers
- Epithelium, respiratory
- Lamina propria
- Lymphoid nodule
- Mucociliary layer
- Muscularis mucosae
- Pulmonary artery
- Pulmonary vessels
- Respiratory bronchioles
- Terminal bronchiole

Respiratory bronchioles
- Adjacent alveoli
- Alveoli
- Associated alveoli
- Capillaries
- Epithelium
- Epithelium, simple squamous
- Pulmonary artery
- Respiratory bronchioles
- Terminal bronchiole
- Alveolar ducts
- Alveolar sac
- Alveoli
- Interalveolar septum
- Knobs
- Macrophages
- Respiratory bronchioles
- Terminal bronchiole

Alveoli
- Air-blood barrier
- Alveolar space
- Capillaries
- Interalveolar septum
- Macrophage
- Monocyte
- Septal cell
- Squamous alveolar cells
- Surfactant

Blood vessels
- Air-blood barrier
- Alveolar macrophages
- Alveoli
- Bronchial (terminal)
- Bronchial blood vessels
- Bronchiole
- Connective tissue septa
- Elastic lamina
- Interalveolar septa
- Macrophages
- Pleura
- Pulmonary artery
- Pulmonary capillaries
- Pulmonary vein
- Respiratory bronchioles
- Pleura
- Pulmonary vein

Pleura
- Alveoli
- Blood vessels
- Connective tissue layer
- Macrophages
- Mesothelium
- Parietal pleura
- Pleura space
- Visceral pleura

CHAPTER

14

LYMPHOID SYSTEM

GENERAL CONCEPTS

- Functions
  - Provides immune surveillance and defense against foreign substances and microorganisms
  - Provides immune tolerance, distinguishing between "self" and "non-self"
  - Absorbs lipids into small lymphoid vessels (lacteals) in intestinal villi for distribution to the blood stream and liver
  - Helps to maintain fluid balance by accumulating tissue fluid and white blood cells in lymph vessels and returning them to the blood
- Overview of lymphoid components
  - *Primary lymphoid organs and structures*
    - *Bone marrow.* Site of origin of T and B lymphocytes. B lymphocytes directly seed secondary lymphoid structures and organs.
    - *Thymus.* T lymphocytes from bone marrow undergo further maturation in the thymus before seeding secondary lymphoid structures and organs.
  - *Secondary lymphoid organs and structures* (from least to most complex)
    - *Diffuse lymphoid tissue*

Digital Histology: An Interactive CD Atlas with Review Text, by Alice S. Pakurar and John W. Bigbee
ISBN 0-471-64982-1

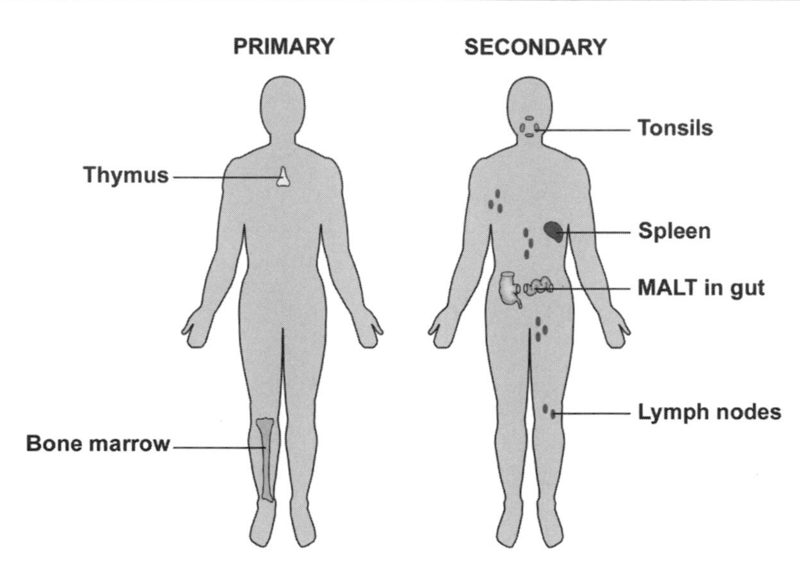

FIGURE 14.1. The primary and secondary lymphoid structures and organs. MALT, mucosal-associated lymphoid tissue.

- *Lymphoid nodules*. Both solitary and in aggregates.
- *Tonsils*
- *Lymph nodes*
- *Spleen*

- Major lymphoid cell types
  - *B lymphocytes* originate and mature in the bone marrow, then seed secondary lymphoid structures and organs. B cells differentiate into B memory cells and plasma cells, providing humoral immunity.
  - *T lymphocytes* originate in bone marrow, mature in the thymus, and subsequently seed secondary lymphoid tissue. T cells differentiate into helper, memory, and cytotoxic cells. T lymphocytes provide cell-mediated immunity and assist B lymphocytes in their humoral response.
  - *Plasma cells* differentiate from B lymphocytes and produce humoral antibodies.
  - *Macrophages* and *dendritic cells* phagocytose foreign matter, enhance the body's response to antigen by "presenting" antigen to lymphocytes, and secrete immunomodulatory factors.
- *Lymph vessels*
  - Are thin-walled vessels lined with endothelium

- Begin as blind-ended lymphatic capillaries in tissues. They accumulate tissue fluid, which is called lymph once it is enclosed by the capillary.
- Gradually increase in diameter and have valves located within their walls. Lymph nodes are positioned along these vessels.
- Unite to form two *lymph ducts* (thoracic and lymphatic ducts) that return lymph to the venous side of the blood vasculature system

- *High endothelial venules (HEVs)*
  - Located in appendix, tonsils, Peyer's patches, and especially in lymph nodes, but not in spleen
  - Endothelium lining these venules is simple cuboidal rather than simple squamous epithelium
  - Allows transport of lymphocytes through the endothelium, thus permitting diapedesis of these cells and the dissemination of immunological information between different regions of the body
- Stroma of lymphoid structures and organs
  - *Reticular cells* produce reticular fibers and act as fixed macrophages as they ensheathe these fibers. This tissue constitutes reticular connective tissue.
  - *Reticular fibers* are composed of collagen type III and form a meshwork that allows fluid to percolate through it while providing delicate, nondistensible support for cells suspended within it.

Components of the Lymphoid System

➢ *Diffuse lymphatic tissue*

- Located in lamina propria of any organ system opening to the exterior of the body, such as respiratory and digestive systems, where an antigen could penetrate the epithelium and enter the lamina propria. Diffuse lymphoid tissue in the lamina propria is part of the mucosal-associated lymphoid tissue (MALT). Diffuse lymphatic tissue is also located in tonsils, lymph nodes, and spleen.
- Composed of an unorganized cluster of lymphocytes and other cells capable of responding to an antigen that reaches it
- Filters and provides immune surveillance for tissue fluid of the lamina propria in which it is located

➢ *Lymphoid nodules*

- Distribution
  - Lamina propria of any organ opening to the exterior of the body. May occur singly (solitary) or in clusters (aggregates) such as in tonsils and Peyer's patches in the small intestine. Lymphoid nodules in the lamina propria are part of MALT.
  - Lymph node and spleen
- Structure
  - *Primary nodule*. The nodule present before antigen stimulation, it consists primarily of densely packed spheres of B lymphocytes.
  - *Secondary nodule*. After antigen stimulation, a central pale core, the *germinal center*, appears. This center is composed of immunoblasts that divide to form lymphocytes that accumulate in the densely packed, peripheral zone.
- Filters and provides immune surveillance for the fluid of the layer/organ in which it is located—tissue fluid in the lamina propria, lymph in lymph nodes, and blood in the spleen; detects specific antigens and causes proliferation of antigen-specific B lymphocytes

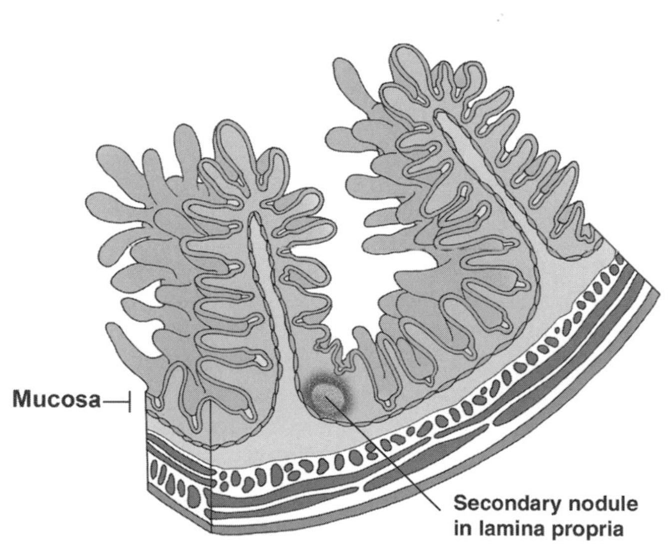

FIGURE 14.2. Nodular lymphoid tissue in the mucosa is part of MALT, mucosal-associated lymphoid tissue. Longitudinal section of the small intestine.

➢ *Tonsils*

- Pharyngeal, lingual, and palatine tonsils are located at the junction of the oral cavity with the oral pharynx and in the nasopharynx.
- Located in the lamina propria of the mucosa
- Structure
  - ◆ Aggregations of lymphoid nodules and diffuse lymphoid tissue
  - ◆ Crypts or folds of surface epithelium invade the tonsils.
  - ◆ Partially encapsulated by connective tissue separating it from underlying tissues
- Filter and provide immune surveillance for the tissue fluid of the lamina propria in which they are located

➢ *Lymph nodes*

- Small, encapsulated, kidney-shaped organs occurring in chains or groups along lymph vessels
- Structure
  - ◆ *Cortex*
    - ■ Capsule of connective tissue surrounds the node and sends short trabeculae into the node. Reticular connective tissue forms the stroma for the remainder of the node.
    - ■ *Outer zone.* Filled primarily with lymphoid nodules composed of B lymphocytes

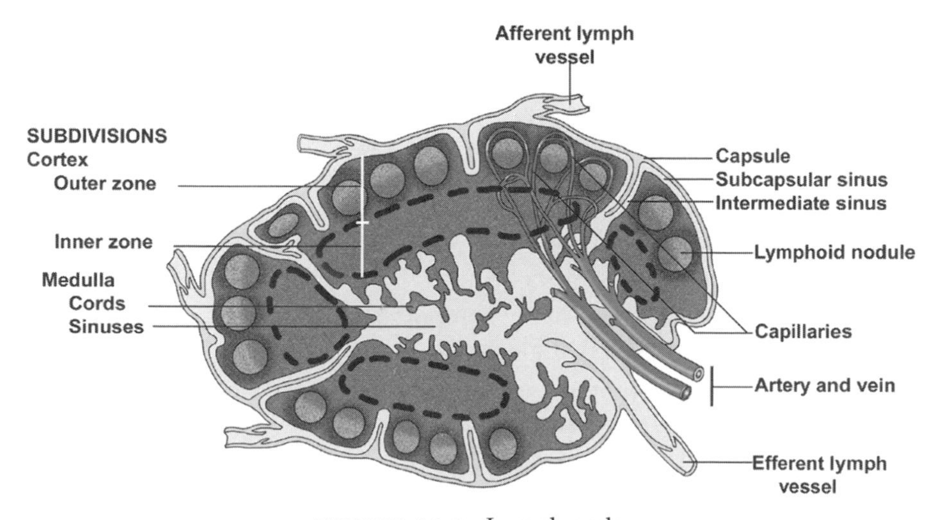

FIGURE 14.3. Lymph node.

- *Inner zone* (paracortex or deep cortex). Filled with diffuse lymphoid tissue composed of T lymphocytes
- Sinuses in cortex. Loose network of macrophages and reticular fibers through which lymph percolates
  - *Subcapsular sinus* lies immediately beneath the capsule and receives incoming lymph fluid from afferent lymphatic vessels that enter through the capsule.
  - *Intermediate sinuses*. Lie adjacent to the trabeculae. Receive lymph from the subcapsular sinus and continue as medullary sinuses
- *Medulla*. Composed of:
  - *Medullary cords* of B lymphocytes that extend from the inner cortex into the medulla
  - *Medullary sinuses*. Continuations of the intermediate sinuses in the cortex. Lymph flows from medullary sinuses into the efferent lymph vessels that exit at the hilum of the node.
- Blood supply. Small arteries enter at the hilum to supply a capillary plexus in the outer cortex. The capillaries anastomose to form HEVs in the paracortex and small veins that exit at the hilum.
- Filter and provide immune surveillance for lymph

➢ *Spleen*

- Encapsulated, intraperitoneal organ located in upper left quadrant of the abdominal cavity
- Structure
  - Capsule surrounds organ, sending trabeculae into the spleen. Larger blood vessels enter through the trabeculae.
  - Subdivisions
    - *White pulp* appears white in fresh specimens and is composed of:
      - *Periarterial lymphoid sheath (PALS)*. A sleeve of T lymphocytes that surrounds a central arteriole as soon as it exits from a trabecula
      - *Lymphoid nodules*, composed of B lymphocytes, are randomly located along and embedded in the PALS.
    - *Red pulp* appears red in fresh specimens because of the abundant venous sinuses it possesses.
      - *Splenic cords* (of *Billroth*). Cords of lymphocytes (T and B), macrophages, plasma cells, and other lymphoid cells suspended in a reticular connective tissue stroma. Surrounded by:

    - *Splenic sinuses.* Venous sinuses separating splenic cords. These sinuses are lined by endothelial cells and surrounded by reticular fibers.
  - The spleen filters and provides immune surveillance for the blood percolating through it. The spleen also phagocytoses aged and abnormal erythrocytes and stores blood.
- Blood flow through the spleen
  - Splenic artery enters at the hilum of the spleen and branches into arteries that lie in the trabeculae.
  - Arteries exit from the trabeculae as *central arterioles* and are immediately surrounded by the PALS. The central arteriole becomes eccentrically located when it is displaced by a lymphoid nodule. Branches from the central arterioles supply the PALS, including forming marginal sinuses at the perimeter of the white pulp.
  - Central arterioles lose their PALS ensheathment and form a series of smaller arterioles in the red pulp. These arterioles either:
    - Open directly into a splenic sinus (*closed circulation*)
    - Open into a splenic cord where the blood percolates through the cells of the cord before entering a splenic sinus (*open circulation*)
  - Trabecular veins are formed by splenic sinuses anastomosing and then entering a trabecula. Trabecular veins anastomose to form the splenic vein.
  - The splenic vein exits at the hilum of the spleen.

➢ *Thymus*

- Thymus is a primary lymphoid organ that receives immature lymphocytes (thymocytes) from the bone marrow. These cells mature in the thymus and are carried to secondary lymphoid structures/organs via the blood vascular system.
- The thymus is located in the superior mediastinum under the sternum. The thymus involutes after puberty.
- Structure
  - A connective tissue capsule surrounds the thymus and extends into the thymus, dividing it into lobules.
  - The stroma is formed by a network of reticular cells of endodermal, rather than the usual mesodermal, origin and are called, therefore, *epithelial reticular cells*. These cells do not form fibers.
  - Each lobule contains an:

  - *Outer cortex* that is densely packed with thymocytes, the developing T lymphocytes. These cells mature in the cortex, then migrate into the medulla where they enter the blood stream for transport to secondary lymphoid structures and organs.
  - *Inner medulla* has fewer thymocytes and, therefore, stains more palely than does the cortex. *Hassall's corpuscles* are the degenerating remains of the epithelial reticular cells with their keratin granules and are diagnostic for the thymus.
- A blood-thymic barrier is formed around capillaries in the cortex, so that the developing lymphocytes are not exposed to circulating antigens.

STRUCTURES IDENTIFIED IN THIS SECTION

Overview
- Artery
- Bone marrow
- Capillary
- Diffuse lymphoid tissue
- Mucosa
  - Epithelial barrier
  - Lamina propria
  - Mucosal glands
  - Muscularis mucosae
- Large intestine
- Lymph capillary
- Lymph nodes
- Lymph vessels
- Lymphoid nodules
- MALT
- Microbes
- Peripheral nerve
- Small intestine
- Spleen
- Thymus
- Tonsils
- Valve
- Vein

Lymphoid tissues
- Adventitia
- Arteriole
- Capsule
- Crypt epithelium
- Crypts
- Dense connective tissue capsule
- Diffuse lymphoid tissue
- Dome
- Endothelium
- Epithelium, stratified squamous
- Germinal center
- High endothelial venules
- Lymph
- Lymphatic vessel
- Lymphoblasts
- Lymphoid nodules
- Macrophages
- Mitotic figures
- Muscularis mucosae
- Nodular lymphoid tissue
- Primary nodule
- Reticular cells
- Secondary lymphoid nodule
- Septa
- Skeletal muscle
- Small lymphocytes
- Smooth muscle
- Solitary lymphoid nodule
- Tonsils
- Valve flap

- Venule

Lymph nodes
- Afferent lymphatics
- Arteriole
- Artery and vein
- Capsule
- Cortex
  - Outer cortex
  - Paracortex
- Cortical sinuses
- Deep cortex
- Efferent lymphatic
- Epithelium, simple squamous
- High endothelial venules
- Hilum
- Lymphocytes
- Lymphoid nodules
- Macrophages
- Medulla
  - Medullary cords
  - Medullary sinuses
- Reticular cells
- Sinuses
- Subcapsular sinus
- Trabeculae
- Valve
- Venule

Spleen
- Capsule
- Endothelial cells
- Macrophages
- Red pulp
  - Red pulp arterioles
  - Splenic cords
  - Splenic sinuses
  - Circulation, closed
  - Circulation, open
- Reticular fibers
- Splenic veins
- Trabeculae
- Trabecular artery
- Venous drainage
- White pulp
  - Central arteriole
  - Germinal centers
  - Lymphoid nodules
  - Marginal zone
  - PALS
  - White pulp vasculature

Thymus
- Blood vessels
- Capsule
- Cortex
- Epithelial reticular cells
- Hassall's corpuscles
- Keratohyaline
- Lobule
- Lymphoblasts
- Medulla
- Septa
- Thymic lymphocytes
- Thymocytes

CHAPTER

15

URINARY SYSTEM

COMPONENTS

- Kidneys. Contain the *uriniferous tubules*, which consist of nephrons and a system of collecting ducts; filter blood and produce urine
- Ureters. Muscular tubes that collect urine output from the kidney and carry it to the urinary bladder
- Urinary bladder. Hollow muscular organ that stores urine
- Urethra. Tube that drains urine from urinary bladder to the exterior

FUNCTIONS OF THE URINARY SYSTEM

- Excretion of waste products of metabolism
- Regulation and maintenance of the fluid volume of the body
- Regulation of acid-base balance
- Regulation of salt concentrations and other compounds in body fluids
- Production of renin, an enzyme that influences blood pressure

MACROSCOPIC ORGANIZATION OF THE KIDNEY

- *Cortex*. Broad outer zone of kidney

Digital Histology: An Interactive CD Atlas with Review Text, by Alice S. Pakurar and John W. Bigbee
ISBN 0-471-64982-1

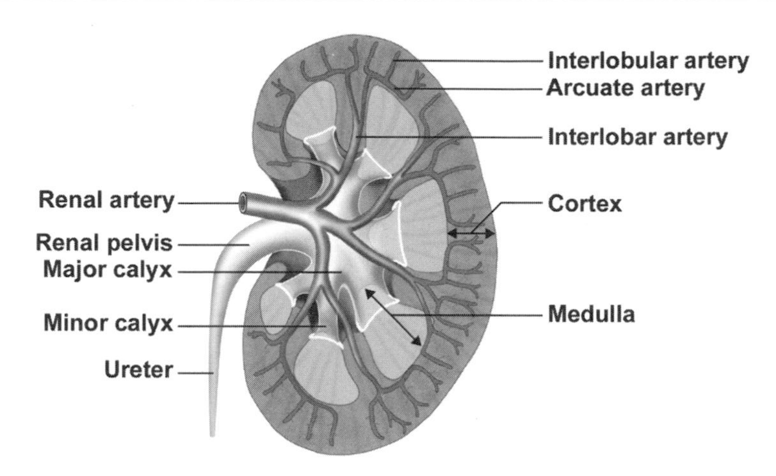

FIGURE 15.1. Extrarenal passageways and vascular supply of the kidney.

- Subdivisions
  - *Labyrinth.* "True" cortical tissue
  - *Medullary rays.* Medullary tissue located in the cortex
- Consists of renal corpuscles, portions of renal tubules, and collecting ducts

➢ *Medulla.* Deep to cortex

- Subdivisions
  - *Renal pyramids.* Inverted cones whose bases are adjacent to the cortex; send "stripes" of medullary tissue into the cortex forming the medullary rays
  - *Renal columns.* Extensions of cortical tissue between renal pyramids
- Consists of portions of renal tubules and collecting ducts

➢ Renal lobulations

- *Renal lobe.* A medullary pyramid, the surrounding renal column extending to the interlobar vessels, and the overlying cortical tissue
- *Renal lobule.* A central medullary ray and the adjacent cortical labyrinth extending to the interlobular vessels

➢ Extrarenal passageways

- *Minor calyx.* Funnel-shaped structure (one for each pyramid) into which the point (apex) of a pyramid projects; urine flows from the pyramid into a minor calyx and several minor calyces unite to form a major calyx.

- *Major calyx.* Four or five per kidney; formed by the confluence of minor calyces
- *Renal pelvis.* Structure formed by the uniting of the major calyces; forms the expanded upper portion of the ureter

THE NEPHRON

➢ 1.5–2 million per kidney

➢ *Renal corpuscle*

- Located in the cortical labyrinth
- Components
  - ◆ *Glomerulus.* A tuft of fenestrated capillaries, whose pores lack diaphragms; filter blood. Formed by an afferent arteriole, the glomerulus indents into Bowman's capsule like a baseball fits into a baseball glove. Blood leaves the glomerulus via the efferent arteriole.
  - ◆ *Bowman's capsule.* Double walled, epithelial capsule with central space called *Bowman's space;* surrounds the glomerulus and receives the fluid filtered from the blood
    - ■ *Parietal layer.* Outer layer, simple squamous epithelium which is reflected at the vascular pole of the renal corpuscle to become the visceral layer; continuous with the proximal tubule at the urinary pole

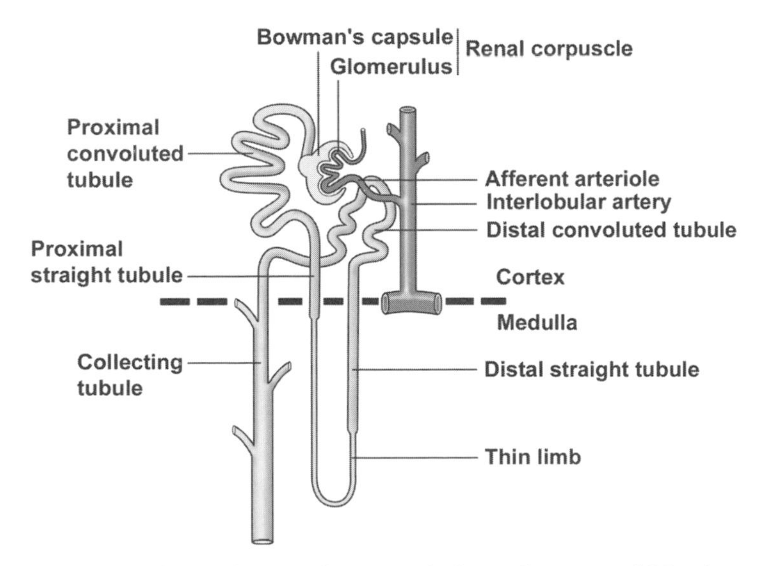

FIGURE 15.2. The nephron, collecting tubule, and associated blood supply.

        - *Visceral layer.* Inner layer surrounding the glomerulus. Consists of a single layer of modified epithelial cells called *podocytes*. The radiating foot processes of these cells give rise to many secondary processes called *pedicels*. Pedicels of adjacent podocytes interdigitate and surround the glomerular capillaries. The slits (*filtration slits*) between the pedicels are bridged by *slit diaphragms*.
    - *Filtration barrier.* Barrier between blood in glomerular capillary and space of Bowman's capsule
        - Fenestrated endothelium of glomerular capillary
        - Thick, fused basal laminae of the podocytes and the glomerular endothelial cells
        - Slit diaphragms between pedicels of visceral layer of epithelium
    - Poles of the glomerulus
        - *Vascular pole.* Where afferent and efferent arterioles enter and leave the renal corpuscle, respectively
        - *Urinary pole.* Where the parietal layer of Bowman's capsule is continuous with the proximal convoluted tubule

➢ Renal tubule

- The glomerular filtrate of the blood continues from Bowman's space into the renal tubule, which meanders first through the cortex, then the medulla, then back to the cortex, and finally enters the collecting duct.
- Regions of the renal tubule
    - Listed in order are regions of the renal tubule through which urine passes
        - Proximal convoluted tubule
        - Proximal straight tubule
        - Thin limbs
        - Distal straight tubule
        - Distal convoluted tubule
    - *Proximal tubule, convoluted portion*
        - Located in labyrinth of cortex; highly convoluted
        - Interconnects parietal epithelium of Bowman's capsule with straight portion of proximal tubule
        - Composed of a simple cuboidal epithelium with microvilli; cells possess numerous infoldings of the basal plasma membrane and many mitochondria

- Absorption of glucose, amino acids, and the majority of salt and water occur here.

- *Loop of Henle.* Located in medullary tissue (i.e., medullary ray and medulla)
  - *Proximal tubule, straight portion (thick descending limb of loop of Henle)*
    - Located either in medullary ray (in cortex) or in medulla
    - Interconnects proximal convoluted tubule with thin limb of Henle's loop
    - Histology is identical to that of the proximal convoluted tubule
    - Absorption of same substances as in proximal convoluted tubule
  - *Thin segment*
    - Found in medulla
    - Interconnects proximal straight tubule with distal straight tubule
    - Frequently makes the "loop" in the loop of Henle
    - Composed of a simple squamous epithelium
    - Actively pumps out chloride, with sodium following passively, to produce a hypertonic urine
  - *Distal tubule, straight portion (thick ascending limb of Henle's loop).*
    - Located either in medulla or in medullary ray (in cortex)
    - Interconnects thin segment with distal convoluted tubule
    - Composed of a simple cuboidal epithelium with inconsistent microvilli. The cytoplasm is less acidophilic and the lumen is wider than in the proximal tubule. The basal plasma membrane is extensively infolded with numerous mitochondria between the folds.

- *Distal tubule, convoluted portion*
  - Located in the labyrinth portion of cortex; highly convoluted
  - Interconnects the distal straight tubule with collecting tubule
  - Histology is identical with the distal straight tubule
  - Returns to a glomerulus to form part of the juxtaglomerular apparatus
  - Major site of salt and water control in the body

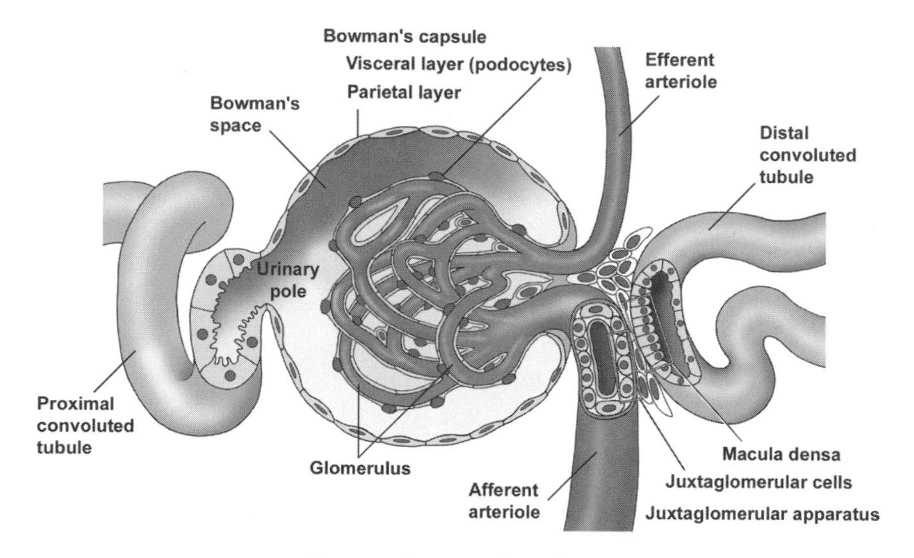

FIGURE 15.3. The renal corpuscle and associated structures.

- *Juxtaglomerular (JG) apparatus*
    - Located at the vascular pole of a nephron; helps regulate blood pressure
    - *JG cells*. Modified smooth muscle cells in wall of an afferent arteriole
    - *Macula densa*. Cluster of modified cells in the wall of a distal convoluted tubule adjacent to the JG cells. The clustering of cells, and therefore of their nuclei, gives the appearance of a "dense spot" in the wall of the distal convoluted tubule.
    - Monitors the tonicity of the urine in the distal tubule. The macula densa affects the adjacent JG cells to adjust their production of *renin*, a hormone that aids in regulating blood pressure.

Excretory Tubules and Ducts and Extrarenal Passages

- Separate embryological origin from the nephron
- Components
    - *Collecting tubule*
        - Composed of simple cuboidal to simple columnar cells; usually displays distinct lateral boundaries between cells

    - Drains urine from the distal convoluted tubule of many nephrons in the cortical labyrinth, enters the medullary ray in the cortex and descends into the medulla
    - Joins with other collecting tubules to form the papillary ducts (of Bellini)
    - Aids in concentrating the urine
- *Papillary ducts.* Located deep in the medullary pyramid near the minor calyces; composed of a tall, pale, simple columnar epithelium. Empty into the minor calyx at the area cribosa at the apex of each pyramid
- *Minor* and *major calyces.* Transport urine to the renal pelvis and into the ureter; lined by *transitional epithelium*
- *Renal pelvis.* Expanded origin of the ureter, lined by transitional epithelium; formed by the union of major calyces
- *Ureter.* Muscular tube connecting the renal pelvis and the urinary bladder, lined by transitional epithelium; two layers of smooth muscle in the upper two-thirds, inner longitudinal and outer circular, with the addition of a third outer longitudinal layer in the lower one-third
- *Urinary bladder.* Lined by a transitional epithelium, a stratified cuboidal epithelium specialized to provide for distension of the organ; a thick muscular wall contains three interlacing layers of smooth muscle.

BLOOD SUPPLY OF THE KIDNEY

➢ *Renal artery.* A branch of the aorta, enters the kidney at the *hilus*; branches to form the interlobular arteries

➢ *Interlobar arteries.* Lie between adjacent pyramids in renal columns and branch into arcuate arteries

➢ *Arcuate arteries.* Arch between medulla and cortex; give rise to interlobular arteries

➢ *Interlobular* arteries. Branch perpendicular to the arcuate artery in the cortex and lie between adjacent lobules; supply a number of afferent arterioles

➢ *Afferent arterioles* supply the glomerulus, entering at the vascular pole of the renal corpuscle

➢ An *efferent arteriole* exits from the glomerulus and forms either *peritubular capillaries*, which nourish the convoluted tubules, or the *vasa recta*. The vasa recta parallel the straight portions of the renal tubule into the medulla and play an important role is concentrating the urine.

STRUCTURES IDENTIFIED IN THIS SECTION

- Blood vessels
  - Afferent arteriole
  - Arcuate vessels
  - Interlobular arteries
  - Peritubular capillaries
  - Vasa recta
- Kidney
  - Cortex
    - Convoluted portion (cortical labyrinth)
    - Medullary rays
  - Medulla
    - Area cribosa
    - Medullary pyramid
    - Minor calyx
    - Renal papilla
- Nephron and collecting ducts
  - Basal lamina
  - Bowman's space
  - Collecting ducts
  - Distal convoluted tubules
  - Endotheial cell
  - Glomerulus
  - Juxtaglomerular apparatus
  - Juxtaglomerular cells
  - Macula densa
  - Medullary collecting duct
  - Mesangial cells
  - Parietal and visceral layers of Bowman's capsule
  - Pedicles of podocytes
  - Podocytes (visceral layer of Bowman's capsule)
  - Proximal convoluted tubules
    - Basal membrane infoldings
    - Brush border (microvilli)
    - Mitochondria
  - Renal corpuscles
  - Straight portions of proximal and distal tubules
  - Thin limb of the loop of Henle
  - Urinary pole
  - Vascular pole
- Ureter
  - Adventitia
  - Smooth muscle layers (muscularis externa)
    - Inner longitudinal
    - Middle circular
    - Outer longitudinal
  - Transitional epithelium
- Urinary bladder
  - Muscularis externa
  - Transitional epithelium with dome cells

CHAPTER

16

ENDOCRINE SYSTEM

GENERAL CONCEPTS

- Unlike exocrine glands, which release their products onto the epithelial surface from which the glands were formed, *endocrine glands* lose contact with their epithelial origin and release their products, called *hormones*, into the extracellular space around the endocrine cells. From here, hormones can affect adjacent cells (*paracrine secretion*) or diffuse into capillaries to be transported in the blood (*endocrine secretion*). Hormones act only on selected cells, called *target cells*, which express specific receptors to mediate the hormone signal.
- The endocrine system consists of organs (pituitary, thyroid, parathyroid, adrenal and pineal glands), clusters of cells (pancreatic islets of Langerhans, theca interna in the ovary and interstitial cells in the testis) and individual cells (enteroendocrine cells in the digestive tract that belong to the belong to the *diffuse neuroendocrine system, DNES*). In addition, numerous organs, which are not exclusively endocrine, also secrete hormones including the kidney, heart, liver, thymus and placenta.
- Endocrine cells and organs have diverse structures, functions and embryological origins. Their hormones can be steroids, (cortisol,

Digital Histology: An Interactive CD Atlas with Review Text, by Alice S. Pakurar and John W. Bigbee
ISBN 0-471-64982-1

testosterone), amino acid derivatives (thyroxine, epinephrine) or peptides and proteins (insulin, growth hormone).

- Endocrine organs are highly vascular and most have fenestrated capillaries which facilitate the entry of the hormone into the blood stream. The cells are usually arranged in plates or cords to maximize surface contact with blood vessels. The organelles of the secretory cells do not show polarity as seen in cells of exocrine glands. Major exceptions to this feature are the follicle cells of the thyroid and individual endocrine cells which are contained in an epithelium, e.g. enteroendocrine cells in the digestive tract.
- Together, the nervous and endocrine systems coordinate functions of all body systems and are functionally integrated as the neuroendocrine system. In fact, the secretory products of some neurons are not neurotransmitters, but rather are neurohormones, because they are released into the blood stream.
- While the nervous and endocrine systems combine to regulate body functions, there are notable differences in the manner in which they do so. Nervous impulses produce their effects within a few milliseconds in contrast to hormones which may require minutes to hours to produce an effect. Furthermore, the effect of a nerve impulse is local whereas hormones often work at a distance and may have diffuse targets.

PITUITARY GLAND (HYPOPHYSIS)

ORIGINS OF THE PITUITARY GLAND

- The pituitary gland consists of two different glands, the adenohypophysis and the neurohypophysis, which are derived embryologically from two distinct tissues.
  - *Adenohypophysis*
    - The adenophypophysis develops from a hollow evagination, *Rathke's pouch,* an outgrowth of stomadeal ectoderm from the roof of the mouth.
    - Rathke's pouch loses its connection with the oral cavity and ascends toward the base of the brain where it contacts the neurohypophysis.
    - Subdivisions
      - *Pars distalis.* Largest subdivision; forms from the anterior wall of Rathke's pouch, constituting >95% of the adenophypophysis

- *Pars tuberalis*. Forms a collar of cells around the infundibulum of the neurohypophysis.
- *Cystic remnants of Rathke's pouch*. Small cysts persisting from the original cavity of Rathke's pouch
- *Pars intermedia*. Forms from the posterior wall of Rathke's pouch at the interface of the adenohypophysis with the pars nervosa of the neurohypophysis; these cells also surround small cystic remnants of Rathke's pouch; this subdivision is rudimentary in humans.

- Neurohypophysis
  - The *neurohypophysis* develops as an outgrowth from the hypothalamus of the diencephalon of the brain, and retains its connection with the brain, abutting the posterior wall of Rathke's pouch.
  - The subdivisions of the neurohypophysis consist of the *infundibulum* and the *pars nervosa*.
- Pituitary terminology

Terminology based on embryonic origin	Pituitary subdivisions	Clinical terminology
Adenohypophysis	Pars distalis	Anterior lobe of pituitary
	Pars tuberalis	
	Pars intermedia	Posterior lobe of pituitary
Neurohypophysis	Pars nervosa	
	Infundibulum	

ADENOHYPOPHYSIS

➢ Cell types

- *Chromophils*
  - *Acidophils*. Hormone-containing granules in the cytoplasm stain with acidic dyes, e.g., eosin
    - *Somatotropes*. Secrete *somatotropin, (growth hormone, GH)* which promotes growth (anabolic)
    - *Mammotropes*. Secrete *prolactin* which stimulates milk production
  - *Basophils*. Hormone-containing granules in the cytoplasm of these cells stain with basic dyes, e.g., hematoxylin

- *Thyrotropes*. Secrete *thyroid stimulating hormone (TSH)* which stimulates thyroid hormone synthesis and release
- *Gonadotropes*. Secrete *luteinizing hormone (LH)* and *follicle stimulating hormone (FSH)*; both hormones are present in males; however, in males, LH can be referred to as *interstitial cell stimulating hormone (ICSH)*; regulate egg and sperm maturation and sex hormone production.
- *Adrenocorticotropes*. Secrete *adrenocorticotropic hormone (ACTH)* which regulates glucocorticoid secretion by adrenal gland

- *Chromophobes*
  - Cells with sparse granule content that do not stain with either hematoxylin or eosin
  - May be degranulated cells or reserve, undifferentiated cells

Hormone(s)	General Cell Type	Specific Cell Type
GH	Acidophil	Somatotrope
Prolactin	Acidophil	Mammotrope
TSH	Basophil	Thyrotrope
FSH/LH	Basophil	Gonadotrope
ACTH	Basophil	Adrenocorticotrope

- Distribution of cell types in the adenohypophysis
  - Pars distalis contains all five cell types
  - Pars tuberalis contains gonadotropes only
  - Pars intermedia contains basophils; however, their function in humans is unclear.
- Regulation of adenohypophyseal secretion
  - Adenohypophyseal hormone secretion is regulated by factors produced by neurons in the hypothalamus. These factors either stimulate or inhibit hormone secretion from their target cells in the adenohypophysis.
  - The *releasing* or *inhibitory factors (neurohormones)* are transported down their axons which terminate in a capillary bed located at the base of the hypothalamus in a region called the *median eminence*. Activity in these neurons causes release of the neurohormones from the terminals and their uptake into the capillaries.

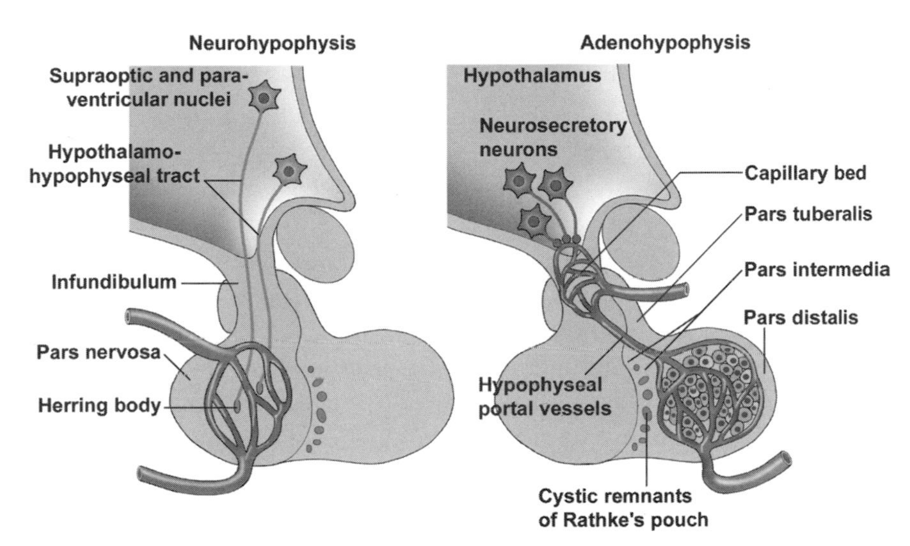

FIGURE 16.1. Comparison of the structure and regulation of secretion of pituitary gland subdivisions.

- The capillaries anastomose into the *hypophyseal portal vessels* which travel down the infundibulum and end in a second capillary network within the adenohypophysis.
- Hypothalamic factors exit this second capillary plexus and either stimulate or inhibit the secretion of hormones from their target acidophil or basophil cells.

NEUROHYPOPHYSIS

➢ Components

- *Infundibulum (hypophyseal stalk)*
  - ◆ Extension from the hypothalamus; continuous with the pars nervosa
  - ◆ Contains the *hypothalamo-hypophyseal tract* which consists of axons from neurons whose cell bodies are located in the *supraoptic* and *paraventricular nuclei* of the hypothalamus
- *Pars nervosa*
  - ◆ Contains axons and axon terminals of the neurons forming the hypothalamo-hypophyseal tract
  - ◆ *Herring bodies.* Expanded axon terminals which accumulate secretory granules containing oxytocin or antidiuretic hormone (vasopressin)

- ■ *Oxytocin* causes smooth muscle and myoepithelial cell contraction.
- ■ *Antidiuretic hormone (ADH)* acts on the kidney tubules to prevent water loss.

◆ Also contains "astrocyte-like" cells, called *pituicytes;* no secretory cells are present.

➢ Regulation of neurohypophyseal secretion

- Oxytocin and vasopressin are synthesized by neurons in the hypothalamus, transported down the axons and stored in axons terminals (Herring bodies) in the pars nervosa.
- Activity in these neurons, in response to physiological signals, causes hormone release (neurosecretion) in a manner similar to release of neurotransmitters.

THYROID GLAND

➢ The thyroid gland consists of two unique structural and functional subdivisions, the thyroid follicles and the parafollicicular cells.

➢ *Thyroid follicles*

- Spheres composed of a single layer of follicle cells; the follicle cells form an epithelium (follicular epithelium) and, thus, these cells have apical and basal surfaces and demonstrate cellular polarity.
- Follicle cells secrete *thyroglobulin,* a glycoprotein that is stored in the center of the follicle.
- Thyroglobulin contains modified tyrosine amino acids that constitute the thyroid hormones, *thyroxine (tetraiodothyronine,* T_4*)* and *triiodothyronine (*T_3*).* Follicle cells take up the stored thyroglobulin and release the hormones into the blood stream.
- Thyroid hormones regulate the basal metabolic rate.

➢ *Parafollicular cells (C cells, clear cells)*

- Occur within the follicular epithelium and in small clusters between follicles
- Possess secretory granules containing the hormone *calcitonin,* which acts to inhibit bone resorption, lowering calcium levels
- Belong to the diffuse neuroendocrine system (DNES)

SYNTHESIS AND RELEASE OF THYROID HORMONES

- Follicle cells synthesize and secrete thyroglobulin from their apical surfaces into the follicle lumen where it is stored. The follicle lumen is an extracellular compartment and, thus, secretion of thyroglobulin constitutes the exocrine secretion of the follicle cells and accounts for the polarity of the cells.
- The tyrosines of thyroglobulin are iodinated in the follicle lumen and rearranged to form the thyroid hormones (T_3 and T_4), which are modified tyrosines that are retained in the primary structure of thyroglobulin.
- The iodinated thyrogobulin is resorbed by pinocytosis into the follicle cells where it is hydrolyzed, liberating T_3 and T_4.
- T_3 and T_4 are released from the basolateral surfaces of the follicle cell and enter the blood stream.
- Active and inactive follicles
  - *Active follicle.* Follicle cells are cuboidal to columnar and are involved with both secretion and resorption of thyroglobulin.
  - *Inactive follicle.* Follicle cells are squamous, reflecting the paucity of secretory organelles and the lack of synthetic and uptake activity.

PARATHYROID GLANDS

- The parathyroid glands are four small, spherical glands that are embedded in the posterior surface of the thyroid gland.
- Cell types
  - *Chief cell*
    - Major cell type, arranged in cords or clumps
    - Small polyhedron-shaped cells with secretory granules visible only with electron microscope
    - Secrete *parathyroid hormone (PTH)* which increases blood calcium levels, primarily by increasing osteoclast activity
  - *Oxyphil cell*
    - Large cell may appear singly or in clumps
    - Heterochromatic nucleus and abundant eosinophilic cytoplasm, due to numerous mitochondria
    - No secretory granules
    - Function is unknown.

ADRENAL GLANDS

STRUCTURE

- Paired glands, each located at the superior pole of a kidney; consist of two distinct subdivisions with different embryological origins
- Subdivisions
    - *Cortex*. Derived from mesoderm and constitutes the major steroid-producing gland
    - *Medulla*. Derived from neural crest and is a major source of epinephrine and norepinephrine neurohormones
- Surrounded by a dense *capsule*

CORTEX

- Features of steroid-secreting cells
    - Abundant smooth endoplasmic reticulum
    - Mitochondria with *tubular cristae* in the zona fasciculata and the zona reticularis; *shelf-like cristae* in the zona glomerulosa
    - Numerous lipid droplets filled with cholesterol, precursor for steroid hormones
    - Secretion is by diffusion, with no hormone storage.
- *Zona glomerulosa*
    - Located immediately beneath the capsule
    - Cells arranged in round clusters
    - Secretes *mineralocorticoids*, e.g., *aldosterone*
- *Zona fasciculata*
    - Middle layer, largest cortical zone
    - Cells arranged in rows perpendicular to the capsule with alternating wide-diameter, fenestrated capillaries
    - Secretes *glucocorticoids* and *androgens*
- *Zona reticularis*
    - Occupies deepest layer of the cortex
    - Cells arranged as anastomosing cords
    - Same secretions as zona fasciculata, *glucocorticoids* and *androgens*

ADRENAL MEDULLA

- Composed of *chromaffin cells*

- Modified adrenergic neurons without axons or dendrites; represent sympathetic ganglion cells
- Polyhedral cells containing abundant dense-core, secretory granules

➢ Chromaffin cells synthesize and release *epinephrine* and *norepinephrine*.

PINEAL GLAND (EPIPHYSIS CEREBRI)

STRUCTURE

➢ Conical-shaped gland, 5–8 mm in length and 3–5 mm in width; develops from the roof of the diencephalon and remains attached by a short pineal stalk

➢ Surrounded by a *capsule* composed of *pia mater*
- Connective tissue septa derived from the pia mater penetrate the gland and subdivide it into indistinct lobules.
- Sympathetic axons and blood vessels enter the gland with the septa.

➢ Cells
- *Pinealocytes*
  - Major cell type, represent modified neurons
  - Euchromatic nucleus, spherical to ovoid, with a prominent nucleolus
  - Cytoplasm not evident with conventional stains; however, silver staining reveals that the cell generally has two or more extensions similar to neuronal processes.
  - Processes end in association with capillaries.
  - Secrete *melatonin*, an indoleamine hormone
- *Interstitial cells*
  - Minor cell type, similar to astrocytes in the brain
  - Nucleus is elongated and more heterochromatic than that of pinealocytes.
  - Possess long processes with intermediate filaments
  - Located among groups of pinealocytes and in the connective tissue septae

➢ *Corpora araneacea (brain sand)*
- Globular, basophilic accumulations of calcium phosphates and carbonates in the interstitial space

- Radio-opaque and, thus, often used as indicators of midline deflection of the brain resulting from pathological conditions

SECRETION

- Major hormone secreted is melatonin which regulates diurnal (circadian) light-dark cycles and seasonal rhythms.
- Melatonin is secreted during darkness; secretion is inhibited by light.
- Retinal stimulation by light is relayed to the pineal via sympathetic innervation from the superior cervical ganglion.

STRUCTURES IDENTIFIED IN THIS SECTION

- Pituitary gland (hypophysis)
  - Adenohypophysis
    - Pars distalis
      - Acidophils
      - Basophils
      - Capillary
      - Chromophobes
    - Pars intermedia
      - Basophils
      - Colloid
      - Remnants of Rathke's pouch
    - Pars tuberalis
      - Basophils
  - Neurohypophysis
    - Infundibulum
    - Pars nervosa
      - Axons
      - Capillary
      - Herring bodies
      - Pituicyte
- Thyroid gland
  - Capillary
  - Colloid
  - Follicle
    - Active follicle
    - Inactive follicle
  - Follicle cells
  - Parafollicular cell (clear cell)
    - Secretory granules
  - Stroma
- Parathyroid gland
  - Capillary
  - Oxyphil cell
  - Principal cell
- Adrenal gland
  - Adrenal cortex
  - Adrenal medulla
  - Capsule
  - Capillaries
  - Chromaffin cell
  - Secretory granule
  - Veins
  - Zona fasciculata
  - Zona glomerulosa
  - Zona reticularis
- Pineal gland
  - Blood vessels
  - Capsule
  - Connective tissue septum
  - Interstitial cell
  - Pia mater
  - Pinealocyte

CHAPTER

17

FEMALE REPRODUCTIVE SYSTEM

COMPONENTS

- Paired ovaries
- Paired oviducts or Fallopian tubes
- Uterus
- Vagina

FUNCTIONS

- Produces female germ cells, *ova* (singular, *ovum*)
- Produces female sex hormones, estrogen, and progesterone
- Receives sperm
- Site of fertilization
- Transports female germ cells, sperm, and conceptus
- Houses and nourish conceptus during pregnancy
- Expels fetus at parturition

Digital Histology: An Interactive CD Atlas with Review Text, by Alice S. Pakurar and John W. Bigbee
ISBN 0-471-64982-1

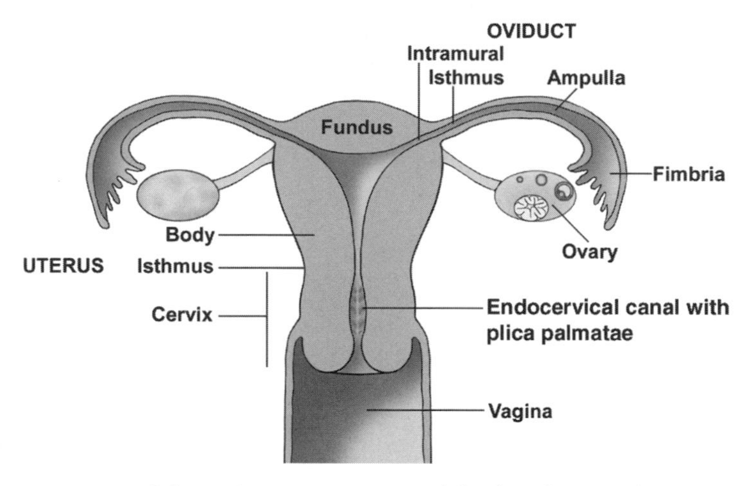

FIGURE 17.1. Schematic representation of the female reproductive system.

OVARY

GENERAL CONSIDERATIONS

- Flattened, ovoid, paired glands
  - Exocrine function. Maturation and release of oocytes, developing female germ cells
  - Endocrine function. Secretion of estrogen and progesterone
- Subdivisions
  - *Cortex*
    - Covered with a serosa
      - *Germinal epithelium.* Simple cuboidal epithelium (mesothelium)
      - *Tunica albuginea.* Underlying, dense connective tissue
    - Contents (exact contents depend on age of the ovary and the stage of the ovarian cycle)
      - *Follicles.* Spheres of epithelial cells surrounding an oocyte. Multiple follicles progress through a series of stages until a single follicle ruptures to release the secondary oocyte at ovulation. Follicles secrete estrogen during the first half of the ovarian cycle.
      - *Corpus luteum.* Formed from the wall of the ovulating follicle after the oocyte is ovulated. The corpus luteum secretes progesterone and estrogen and is present during the second half of the ovarian cycle.
      - *Atretic follicles.* Degenerating follicles that are not ovulated
      - *Corpus albicans.* Degenerating corpus luteum

- *Medulla*. Inner region composed of connective tissue, blood vessels, nerves

Oogenesis

- Oogenesis is the process by which a diploid somatic cell, an oogonium, in the fetal ovary becomes a haploid ovum in the adult after fertilization occurs.
- Stages
  - *Oogonia* in the fetal ovary divide mitotically to form diploid, primary oocytes that are located in primordial follicles. Unlike spermatogonia, oogonia do not replenish their own cell line.
  - *Primary oocytes* immediately begin the first meiotic division, which arrests in prophase. Primordial follicles, each containing a primary oocyte, are the only follicles present from birth until puberty, when selected follicles go through a series of changes during each ovarian cycle, resulting in ovulation.
  - A *secondary oocyte* is formed during the hours preceding ovulation in each ovarian cycle. A primary oocyte in a Graafian follicle completes meiosis I to form a haploid, secondary oocyte. This secondary oocyte, the oocyte that is ovulated, begins meiosis II but arrests in metaphase.
  - An *ovum*, the mature, haploid germ cell, is formed only if fertilization occurs.

Follicles

- *Primordial follicle*
  - Contains a *primary oocyte*
  - *Follicular cells* form a simple squamous epithelium around the oocyte.
  - Is the only follicle present until puberty
- Primary follicles
  - *Primary unilaminar follicle*
    - Contains a primary oocyte
    - Follicular cells form a simple cuboidal or columnar epithelium around the oocyte.
  - *Primary multilaminar follicle*
    - Contains a primary oocyte
    - Follicular cells form a stratified epithelium around the oocyte.

- *Zona pellucida*, formed by both the oocyte and adjacent follicular cells, is a thick glycoprotein band surrounding the oocyte.
- *Theca folliculi*, a layer located outside the basement membrane of the follicular cells, is formed by the differentiation of the surrounding multipotential stromal cells.

➢ *Secondary follicle*

- Contains a primary oocyte
- Follicle cells increase in size and number and produce a follicular liquid.
- *Follicular liquid* accumulates in antral spaces between follicular cells. Multiple antral spaces eventually coalesce to form a single antrum.
- The *granulosa layer* (*granulosa cells or stratum granulosum)* are follicular cells surrounding the antrum. These cells convert androgens, produced in theca interna, into estrogen.
- The *cumulus oophorus* is a hillock of granulosa cells in which the primary oocyte is embedded. The innermost layer of cumulus cells, immediately surrounding the oocyte, forms the *corona radiata*.
- Theca folliculi develops into:
  - *Theca interna*, located outside the basement membrane of the follicular cells, is composed of cells that secrete the steroid hormone androgen.
  - *Theca externa*, composed of multipotential connective tissue cells, resembles a layer of flattened fibroblasts. The theca externa serves as a reserve cell source for the theca interna.
- Usually only a single secondary follicle progresses to the mature follicle stage.
- Follicular growth and maturation is influenced by follicle stimulating hormone (FSH), secreted by the pituitary gland, and estrogen, aromatized by granulosa cells from androgen produced by the theca interna.

➢ *Mature* (*Graafian*) *follicle*. The follicle that will rupture, ovulating a secondary oocyte. Present only during the day preceding ovulation. Changes occurring during the time it is present include:

- Increase in follicular liquid that greatly increases antral and follicle size; follicle will reach a diameter of ~2.0 cm.
- Granulosa and theca interna cells begin formation of *corpus luteum*.
- Enlarged follicle bulges from the ovarian surface, thinning the ovarian tissue covering the follicle and forming a *stigma*.

  - Oocyte and surrounding cumulus oophorus detach from the granulosa layer and lie free in the antral space.
  - Meiosis I is completed with the formation of a *secondary oocyte* and first polar body. Meiosis II arrests in metaphase.
  - *Ovulation.* Day 14 of ovarian cycle
    - The Graafian follicle ruptures at the stigma, releasing the haploid secondary oocyte, cumulus oophorus, follicular liquid, and blood.
    - Oocyte and the surrounding cumulus are transported through the oviduct to the ampulla to await fertilization. Fertilization triggers the completion of meiosis II and the formation of an *ovum.*
    - The follicle wall continues its conversion to a corpus luteum.
    - Ovulation is stimulated by a surge of luteinizing hormone (LH) from the pituitary gland.
- *Atretic follicles*
  - The process of oocyte/follicular atresia begins before birth and continues throughout the life of a woman. Of the ~2 million primordial follicles and their primary oocytes present at birth, only about 450,000 oocytes/follicles remain at puberty and about 450 of those will be ovulated. The remainder degenerate, thereby producing more atretic than "normal" follicles.
  - Atresia can occur at any stage of follicular development and will begin in different layers of the follicle or oocyte depending on the follicle's stage of development. Therefore, many varieties of atretic follicles can be seen.

Corpus Luteum

- The *corpus luteum* is large, spherical, infolded body functional during the second half of the ovarian cycle.
- Functional stage
  - The corpus luteum is formed by differentiation of the granulosa and theca interna cells in the Graafian follicle before and after ovulation.
  - Its formation is stimulated by luteinizing hormone (LH) secreted by the pituitary gland.
  - The life span of the corpus luteum is finite, lasting about 12 days during the average cycle, during days 14–26.

- Composed of
  - *Granulosa lutein cells*. Form from cells in the granulosa layer; typical steroid-secreting cells; major component of the corpus luteum
  - *Theca lutein cells*. Form from theca interna cells; typical steroid-secreting cells but smaller than granulosa lutein cells; remain at the outer boundary of the corpus luteum surrounding the granulosa lutein cells; form a peripheral layer and the infoldings of the corpus luteum
- Secretes progesterone and estrogen
- If pregnancy occurs, placental hormones maintain the corpus luteum, and it is known as the corpus luteum of pregnancy. This structure is functional for the first trimester of pregnancy.

➢ Degenerating stage. *Corpus albicans*

- Consists of a white mass of scar tissue composed of much collagenous material and scattered fibroblasts
- Results from the degeneration of the corpus luteum

CYCLICITY OF OVARY—BASED ON AN AVERAGE 28-DAY CYCLE

➢ *Follicular phase* (days 1–13). Follicles are differentiating and secreting estrogen. Follicles are developing while menstruation (days 1–4) is occurring.

➢ *Ovulation* (day 14). Graafian follicle ruptures, releasing secondary oocyte.

➢ *Luteal phase* (days 15–28). Corpus luteum is the functional ovarian structure, secreting progesterone and estrogen. Hormone secretion diminishes after day 26.

OVIDUCT

SUBDIVISIONS

➢ The oviducts are paired, 12-cm-long tubes that have four subdivisions.

- *Infundibulum*. Funnel-shaped, free end with finger-like fimbria embracing the ovary
- *Ampulla*. Thin walled, lateral two-thirds; fertilization occurs here near its junction with the isthmus
- *Isthmus*. Thicker walled, medial one-third

- *Intramural* (interstitial). Within uterine wall; lumen is continuous with uterine lumen.

STRUCTURE

- *Mucosa*
  - Shows gradations from infundibulum to intramural subdivisions
  - Exhibits complex mucosal folds that are most elaborate in the infundibulum and are sparse in the intramural subdivision
  - Epithelium. *Simple columnar* composed of *ciliated cells*, that are most abundant in the infundibulum, and *secretory cells*, that are most abundant in the intramural portion
  - Muscularis mucosae is lacking.
- *Submucosa* is continuous with lamina propria, forming a continuous connective tissue layer.
- *Muscularis externa* has poorly defined inner circular and outer longitudinal smooth muscle layers, which are thinnest in the infundibulum and thickest in the intramural portion.
- *Serosa*. Covers the outer surface except in the intramural portion

UTERUS

GROSS ANATOMY

- A single, pear-shaped, and pear-sized organ
- Subdivisions
  - *Fundus*. Domed portion above entrance of oviducts
  - *Corpus* or *body*. Major portion of the uterus
  - *Isthmus*. Constricted portion at junction of cervix and body
  - *Cervix*. Located above and within the vagina, defining supravaginal and vaginal portions

GENERAL HISTOLOGICAL ORGANIZATION OF THE BODY AND FUNDUS

- *Perimetrium*. Outermost layer of serosa, covering the upper and posterior regions only; an adventitia surrounds the remaining portions that lie adjacent to the urinary bladder.
- *Myometrium*. A thick, well-vascularized band of smooth muscle that is arranged in ill-defined layers. The myometrium forms the major portion of the uterus and is equivalent to a muscularis externa.

- *Endometrium*. Mucosa.
  - Components
    - Epithelium is *simple columnar*, some with cilia.
    - *Lamina propria* (endometrial stroma) contains multipotential (stromal) cells and abundant ground substance.
    - *Simple tubular glands*
  - Zonation
    - *Functional zone* (stratum functionalis). Luminal two-thirds that is sloughed during menstruation
    - *Basal zone* (stratum basalis). Firmly attached to the myometrium and retained during menstruation. Cell growth from this zone restores functional zone following menstruation.

ARTERIAL SUPPLY TO ENDOMETRIUM

- *Basal (straight) arteries*. Remain in and supply basal zone
- *Spiral arteries*. Located at the junction of the basal and functional zones, spiral arteries extend into and supply the functional zone and a capillary plexus beneath its surface epithelium. Early in the menstrual cycle, spiral arteries are nearly straight, but they become highly coiled later in the cycle.
- A capillary plexus lies under the surface epithelium.

ENDOMETRIAL CHANGES IN THE BODY AND FUNDUS DURING MENSTRUAL CYCLE

- Coordinated with ovarian cycle and controlled by its hormones; approximately 28 days long
- Phases of uterine menstrual cycle
  - *Menstrual phase*. Days 1–4, the clinical beginning of the cycle; however, this phase actually marks the end of the cycle. At the end of menstruation the functional zone has been sloughed and only the basal zone remains.
  - *Proliferative phase* (estrogenic). Days 5–14
    - Ovarian follicles are growing and secreting estrogen.
    - The functional zone proliferates and regenerates from the basal zone.
      - Proliferation of epithelial and stromal cells thickens the endometrium.

    - Glands are initially straight but become slightly wavy toward the end of the phase.
    - Spiral arteries grow with endometrium but are difficult to see because they are not yet coiled (i.e., are straight).
- *Secretory phase* (luteal). Days 15–26
  - Corpus luteum is present and functional.
  - Uterine glands are actively secreting glycogens and glycoproteins by day 20 or 21 when implantation could occur.
  - Endometrial changes leading up to day of implantation
    - Glands enlarge and become tortuous, coiled, and secretory.
    - Spiral arteries lengthen, are highly coiled, and readily visible.
    - Stromal cells become *(pre)decidual cells* (about day 24).
      - Large, pale cells, with glycogen and lipid, located under the surface epithelium and around spiral arteries
      - If pregnancy occurs, these cells are called *decidual cells* and form the decidua, the maternal placenta
      - If pregnancy does not occur, these same cells are called predecidual cells and they are sloughed with menses.
- *Premenstrual phase*. (ischemic portion of secretory phase) Days 26–28
  - Estrogen and progesterone secretion from the ovarian corpus luteum decreases.
  - Compression of the endometrium, resulting from lack of hormones from the corpus luteum, causes:
    - Constriction of the spiral arteries, which results in ischemia in the overlying tissue in the functional zone.
    - The ischemia causes the endometrium to become necrotic and disrupted.
    - Spiral arteries reopen and blood flows into the ischemic tissue, resulting in bleeding from the spiral arteries into the stroma.
    - Cycles of compression and reopening of the arteries leads to degeneration of the functional zone and menstruation.
- *Menstrual phase*. Days 1–4
  - The functional zone becomes necrotic and is sloughed as menses.
  - Menstrual flow contains blood, tissue fragments, and uterine fluids.
  - Only the basal zone remains, from which the functional zone will be regenerated.

CERVIX

- The cervix is the lowest portion of the uterus, beginning above and extending into the vagina.
- *Endocervix*. Surrounds the *(endo)cervical canal*
  - Structure
    - Mucosa
      - *Simple columnar epithelium* with cilia and many mucus-secreting cells
      - Lamina propria is filled with epithelial folds, the *plica palmatae*, that are lined with mucus-secreting cells.
      - Nabothian cysts occur when a fold becomes occluded.
    - Remainder of cervix consists of connective tissue with some smooth muscle.
  - Cyclic changes and functions of the cervix
    - Mucosa is not sloughed during menstruation. Spiral arteries are absent.
    - Cyclic changes in the cervical mucus
      - At mid-cycle, secretions are abundant and the molecules are linearly arranged, facilitating the movement of sperm through the cervix and/or into the plica palmatae for storage. The alkalinity of the cervical mucus neutralizes the low vaginal pH, providing a more favorable environment for spermatozoa.
      - At other times during the cycle, cervical mucus is more viscous, making sperm penetration difficult.
- *Ectocervix*
  - Portion of the cervix that protrudes into the vagina
  - Covered by *moist stratified squamous epithelium*. The junction of this epithelium with the simple columnar epithelium of the endocervical canal is abrupt and is called the *external os*.

VAGINA

STRUCTURE

- *Mucosa*
  - Epithelium. *Stratified squamous nonkeratinized* that accumulates glycogen; no glands are present.

- Lamina propria. Rich with numerous blood vessels and elastic fibers.
- Muscularis mucosae is lacking.
- Rugae (folds) allow for expansion.

➢ *Submucosa* is continuous with the lamina propria, forming a single connective tissue layer.

➢ *Muscle layer*

- Inner circular and outer longitudinal smooth muscle layers intertwine.
- Skeletal muscle surrounds the vaginal orifice.

➢ *Adventitia* of loose connective tissue

Cyclic Changes

➢ The epithelium synthesizes and accumulates glycogen, becoming thick and proliferative by midcycle.

➢ Glycogen use by bacteria produces lactic acid, generated by the breakdown of the glycogen. The lactic acid creates an acidic environment, particularly near ovulation when the glycogen is most abundant.

Structures Identified in This Section

Ovary
- Antral spaces
- Antrum
- Atretic follicles
- Basement membrane
- Corpus albicans
- Corpus luteum
- Cortex
- Cumulus oophorus
- Follicles
- Follicular cells
- Germinal epithelium
- Granulosa lutein cells
- Granulosal layer
- Medulla
- Primary multilaminar follicle
- Primary oocytes
- Primary unilaminar follicle
- Primordial follicle
- Secondary follicle
- Stromal cells
- Theca externa
- Theca folliculi
- Theca interna
- Theca lutein cells
- Tunica albuginea
- Zona pellucida

Oviduct
- Basal bodies
- Cilia
- Lamina propria
- Mucosa
- Muscularis externa
- Oviduct
- Secretory cells
- Simple columnar epithelium

Uterus
- Basal zone
- Blood (extravasated)

Cervix
Degeneration
Ectocervix
Endocervical canal
Endometrium
Functional zone
Menstrual stage
Mucous secretions
Myometrium
Plicae palmatae
Predecidual cells
Premenstrual stage
Proliferative stage
Secretory stage
Simple columnar epithelium
Simple tubular glands
Spiral artery
Stratified squamous moist epithelium

Vagina
Anterior fornix
Blood vessels
Connective tissue
Glycogen
Muscularis externa
Posterior fornix
Stratified squamous moist epithelium
Vein

PLACENTA

GENERAL CONSIDERATIONS

- Definition. An apposition or fusion of membranes of the fetus (chorion) with maternal uterine mucosal layers (the decidua) to produce hormones and to exchange gases and nutrients
- Function
  - Provides exchange of respiratory gases between maternal and fetal circulations
  - Provides nutrients for and removes wastes from the conceptus
  - Secretes hormones
  - Transports some macromolecular materials (e.g., viruses, IgG, alcohol)

BLASTOCYST

- A *blastocyst* is the stage of the embryo that implants into the uterus about day 20 or 21 of the menstrual cycle.
- Composition
  - *Trophoblast cells* form the peripheral rim of the fluid-filled blastocyst cavity; these cells will form the fetal portion of the placenta.
  - *Inner cell mass*, an eccentrically located cluster of cells inside the trophoblast at one pole of the blastocyst, develops into the embryo.

CHORION, THE FETAL PLACENTA

➢ The *chorion* is composed of extraembryonic connective tissue and two cell layers derived from the trophoblast, called the cytotrophoblast and the syncytiotrophoblast.

➢ Formation of the fetal placenta
- Forms within the 23 days following ovulation
- *Trophoblast* erodes into the maternal endometrium at implantation and immediately differentiates into:
  - *Cytotrophoblast*. Inner (toward embryo) single layer that gives rise to the syncytiotrohoblast
  - *Syncytiotrophoblast*. Multinucleated, outer syncytium formed from the cytotrophoblast; aggressively invades the uterine endometrium
- *Lacunae*, separated by columns of syncytiotrophoblast, develop in the syncytiotrophoblast and coalesce, becoming filled with maternal blood from the spiral arteries. After the formation of villi, lacunae are called *intervillous spaces*.
- *Tertiary (definitive) chorionic villi*
  - The final in a series of villi that protrude from the syncytiotrophoblast columns into the lacunae, increasing the surface area of the placenta that is exposed to maternal blood.
  - Formed by invasion of syncytiotrophoblast columns by cytotrophoblastic cells and then by fetal connective tissue. Finally, fetal blood vessels invade into and form within the fetal connective tissue.
  - Composition. From internal to external
    - Core of *fetal connective tissue* with fetal blood vessels
    - *Cytotrophoblast*
      - Large, discrete cells with large, euchromatic nuclei form a single layer around the connective tissue core.
      - Changes during pregnancy
        - Early pregnancy. Cytotrophoblast forms a continuous layer beneath the syncytiotrophoblast.
        - Late pregnancy. Cytotrophoblast layer thins and is even lacking in some areas, thus decreasing the thickness of the barrier through which nutrients/wastes must pass.
    - *Syncytiotrophoblast*
      - Covers outer surface of the villus, facing the intervillous spaces

- Composed of a single cell (syncytium), possessing multiple nuclei; formed by fusion of cytotrophoblastic cells
- Possesses microvilli and abundant organelles associated with both protein and steroid hormone production
- Functions
  - Forms part of the interhemal barrier (barrier between fetal and maternal blood vessels)
  - Secretes a variety of hormones, such as human chorionic gonadotrophin (HCG), human placental lactogen (HPL, somatomammotropin), human placental thyrotropin (HPT), and estrogen and progesterone

- Alterations during pregnancy
  - Early pregnancy. Villi are thick, with a few, thick branches.
  - Late pregnancy. Villi are much more slender, with profuse branching, thus increasing surface area.

- Trophoblastic shell
  - Outer rind of the fetal placenta abutting the maternal decidua
  - Composed of cytotrophoblast positioned between the syncytiotrophoblast and the external maternal decidua.
  - Formed by cytotrophoblast cells growing through and then spreading out beneath the syncytiotrophoblast columns

THE DECIDUA, THE MATERNAL PLACENTA

- The *decidua* is formed from endometrial stromal cells that differentiate into decidual cells beginning about day 24 of the menstrual cycle (about 3 days post implantation).
- Partitioned into three subdivisions, named according to the position of each in relation to the developing conceptus
  - *Decidua basalis.* Underlies the implanted conceptus (beneath trophoblastic shell), forming the maternal portion of the functional placenta
  - *Decidua capsularis.* Covers the luminal surface of the conceptus, separating it from the uterine lumen; will eventually fuse with decidua parietalis of the opposite side, obliterating the uterine lumen
  - *Decidua parietalis* lines the remainder of the uterus.

Maternal Cotyledons

- *Maternal cotyledons* form a *"lobe"* of the placenta, easily identifiable on the maternal surface of the placenta as domed-shaped protrusions from the placenta.
- Up to 35 lobes are present.
- Malformations (discrepancies) of these cotyledons, when examined at delivery, can indicate fetal abnormalities.

Placental Blood Flow

- Fetal
  - *Umbilical arteries* are two in number and travel from the fetus through the umbilical cord to the placenta, carrying blood that is high in carbon dioxide and low in nutrient content. Umbilical arteries branch into capillaries within the tertiary villi.
  - Fetal capillaries are located in the tertiary chorionic villi (definitive villi).
  - Fetal veins from the capillaries anastomose to form the single *umbilical vein* that returns oxygen-rich, nutrient-rich blood to the conceptus.
- Maternal
  - *Spiral arteries* penetrate trophoblastic shell and spurt blood into intervillous spaces, bathing villi.
  - Branches of uterine veins carry blood away from the intervillous spaces.

Placental Interhemal Membrane

- The *placental interhemal membrane* separates maternal and fetal blood supplies, which normally do not mix.
- Beginning in the fetal capillary, this barrier consists of:
  - Capillary endothelial cell (of villus) and its basement membrane
  - Fetal connective tissue of villus (usually lacking in late pregnancy)
  - Cytotrophoblast (may be lacking in late pregnancy) and its basement membrane
  - Syncytiotrophoblast

STRUCTURES IDENTIFIED IN THIS SECTION

Anchoring villi
Chorion
Cytotrophoblast
Decidua basalis
Decidual cells
Fetal capillaries
Fibrinoid
Intervillous space
Mucoid (fetal) connective tissue
Placental hemal barrier
Syncytiotrophoblast
Tertiary chorionic villus
Trophoblastic shell
Umbilical vessels

BREAST

ORGANIZATION OF THE BREAST

- Each breast is a collection of 15–20 separate *mammary glands*, which are modified sweat glands.
- Each gland or lobe of the breast is further subdivided into lobules.
- Each gland in its function state is classified as a *compound tubulo-alveolar gland*.
- Each gland has its own *lactiferous duct*, which empties at the nipple.

ORGANIZATION OF MAMMARY GLANDS

- *Stroma*. Connective tissue framework of the breast
  - *Interlobular connective tissue*, composed of dense, irregular connective tissue with abundant adipose tissue, separates lobules. This tissue is sparsely cellular, containing mostly fibroblasts and adipocytes.
  - *Intralobular connective tissue* is composed of loose connective tissue and lies within lobules and surrounds ducts and alveoli (parenchyma) of the gland. This connective tissue is highly cellular, containing many plasma cells, lymphocytes, and macrophages, as well as fibroblasts.
- *Parenchyma*. Functional components of the breast
  - *Ducts*. Form the majority of an inactive gland and are always present. Consist of an epithelial lining, which can be secretory, and *myoepithelial cells*. Development of the ducts is regulated by estrogen.
  - *Alveoli*. Derived from outgrowths of the ducts and are only present during later stages of pregnancy and lactation. Alveoli consist of alveolar cells and myoepithelial cells. *Alveolar cells* are the major

cells responsible for the synthesis and secretion of milk, and their development is regulated by progesterone.

FUNCTIONAL STAGES OF THE BREAST

- *Prepuberty*. Composed entirely of the duct system; no secretory alveolar units are present. At this stage, the breast in the male and female are similar.
- *Puberty (female)*
  - Enlargement of the breast is due primarily to the accumulation of adipose tissue.
  - Rising estrogen levels at this time stimulates the growth and branching of the duct system.
  - No secretory alveoli are present.
- *Inactive (nonpregnant)*. Minor alveolar development with a slight amount of secretory activity and fluid accumulation may occur during mid to late phases of the menstrual cycle.
- *Pregnancy*
  - Early to mid pregnancy
    - Prominent increase in duct branching induced by estrogen; development of alveoli as evaginations from those ducts is induced by progesterone.
    - Interlobular connective tissue becomes more cellular with increased numbers of plasma cells, polymorphonuclear leukocytes, and lymphocytes.
  - Late pregnancy
    - Significant breast enlargement due to hypertrophy of alveolar cells
    - Lumens of ducts and alveoli widen as secretory products accumulate
- *Lactation*
  - Alveolar cells secrete by both merocrine and apocrine modes of secretion.
    - *Apocrine*. Lipid secretion; lipid droplets coalesce in the apical cytoplasm and are released along with some membrane and surrounding cytoplasm.
    - *Merocrine*. Protein secretion; protein, packaged in membrane-bound, secretory vesicles in the Golgi, are released by exocytosis.

- Two types of secretory products
  - *Colostrum*. Secreted for the first few days after birth; protein rich with a high antibody content.
  - *Milk*. Secretory product released after the colostrum phase; milk has a high lipid content compared with colostrum and also contains protein, carbohydrates, and antibodies.
- Secretion and ejection of milk is maintained by a *neurohormonal reflex arc*. Infant suckling stimulates sensory nerves, whose activity results in prolactin release from the pituitary gland to maintain milk synthesis and secretion. Similarly, suckling causes oxytocin release, which stimulates the ejection of milk due to contraction of myoepithelial cells.
- Cessation of suckling results in decreased secretory activity, degeneration of alveoli, and the end of lactation.

STRUCTURES IDENTIFIED IN THIS SECTION

Stages of the breast
- Inactive
- Pregnant
- Lactating

Structures in the breast
- Adipose connective tissue
- Alveolus (acinus)

Ducts
- Intralobular
- Interlobular

Lobules
- Interlobular connective tissue
- Intralobular connective tissue

Tubules

CHAPTER

18

MALE REPRODUCTIVE SYSTEM

GENERAL CONCEPTS

- Components
  - Testis. Paired organs
    - Seminiferous tubules
    - Rete testis
  - Genital ducts
    - Epididymis. Paired organs
    - Ductus deferens. Paired ducts
    - Ejaculatory duct. Paired ducts
    - Urethra
  - Major genital glands
    - Seminal vesicles. Paired glands
    - Prostate. Single gland
    - Bulbourethral glands. Paired glands
- Functions
  - Produce sperm
  - Produce male sex hormones

Digital Histology: An Interactive CD Atlas with Review Text, by Alice S. Pakurar and John W. Bigbee
ISBN 0-471-64982-1

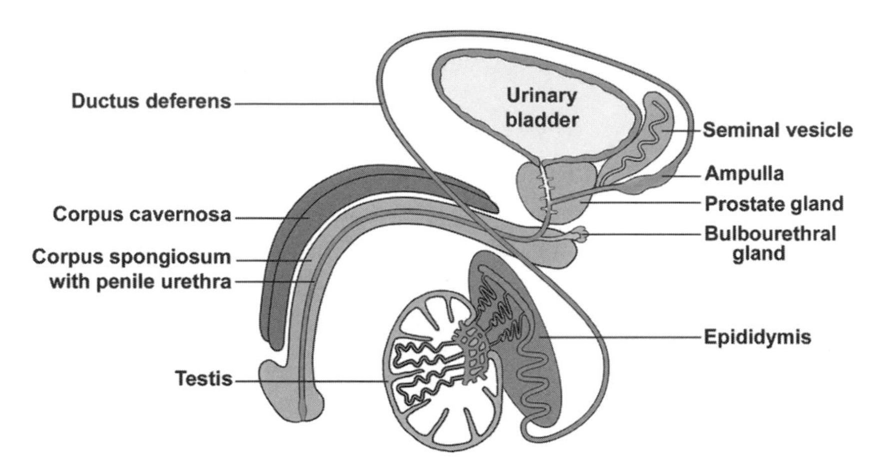

FIGURE 18.1. Schematic representation of the male reproductive system.

- Produce seminal fluid
- Propel sperm and seminal fluid (semen) to exterior

TESTIS

GENERAL ORGANIZATION

- Paired, ovoid organs; serve both exocrine (sperm production) and endocrine (testosterone production) functions; suspended in the scrotum
- Coverings and connective tissue framework (from external to internal):
  - *Tunica vaginalis.* A serosa (peritoneum) that accompanied the testis embryologically in its retroperitoneal descent from the abdomen to the scrotum. Covers the anterior and lateral surfaces of the testes but not their posterior surfaces.
  - *Tunica albuginea.* A layer of dense connective tissue beneath the tunica vaginalis, encapsulating the seminiferous tubules
  - *Mediastinum* of the testis
    - Thickening of tunica albuginea, projecting into the testis from its posterior surface
    - Seminiferous tubules converge at the mediastinum where they join the rete testis.
    - *Rete testis.* Interconnecting channels in the mediastinum that receive the contents of the seminiferous tubules

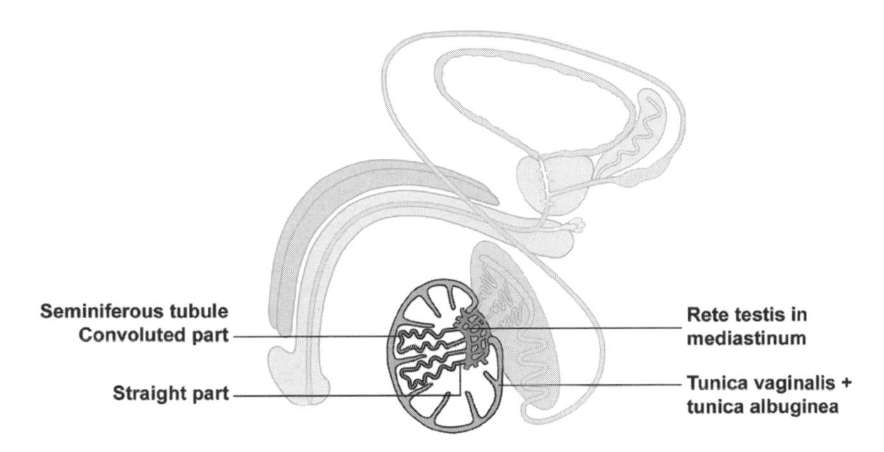

FIGURE 18.2. Components of the testis.

- *Septa*, connective tissue partitions extending from the mediastinum to the tunica albuginea, separate each testis into about 250 lobules.

➢ Internal structure of testis

- Consists of lobules that are pyramidal in shape with their apices directed toward the mediastinum and their bases adjacent to the tunica albuginea
- Composition of lobules
  - ◆ Stroma. Loose connective tissue, many blood vessels and lymphatics
  - ◆ Parenchyma
    - ■ *Seminiferous tubules, convoluted portions*
      - One to four loop-shaped, tortuous tubules per lobule with both ends opening at the mediastinum
      - Composed of seminiferous epithelium where spermatozoa production occurs
      - Surrounded by a *tunica propria*, a connective tissue layer located beneath the basal lamina of the seminiferous epithelium. Myoid cells, possessing contractile properties, are located in this layer.
    - ■ *Seminiferous tubules, straight portions* (tubuli recti) are located at the periphery of the mediastinum. Interconnect the convoluted portions of the seminiferous tubule with the rete testis in the mediastinum
    - ■ *Interstitial cells of Leydig*. Clusters of endocrine-secreting cells lie outside the seminiferous tubules within the CT stroma; produce testosterone

MICROSCOPIC APPEARANCE OF THE PARENCHYMA OF THE TESTIS

- ➢ Endocrine portion. *Interstitial cells of Leydig*
  - Arranged as clusters of cells in the stroma between seminiferous tubules
  - Cytology
    - ◆ Euchromatic nucleus
    - ◆ Eosinophilic cytoplasm possesses cytological features of steroid-producing cells, such as extensive SER, large numbers of lipid droplets and mitochondria with tubular cristae.
  - Function. Secrete testosterone under the influence of luteinizing hormone (LH).
- ➢ Exocrine portion. *Seminiferous*, or *germinal, epithelium* lining the convoluted portions of the seminiferous tubules
  - *Supporting cells of Sertoli*
    - ◆ Tall, columnar cells that rest on the basal lamina and extend to the lumen
    - ◆ Nucleus is euchromatic, ovoid, and infolded; its long axis usually lies perpendicular to, but not immediately adjacent to, the basement membrane.

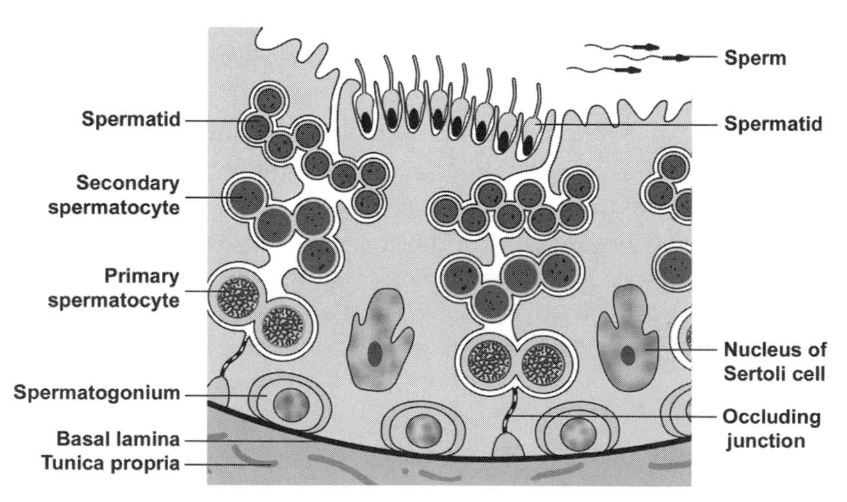

FIGURE 18.3. Seminiferous epithelium in the convoluted portion of the seminiferous tubule.

- Numerous lateral processes surround and invest the maturing germ cells. The most basal of these processes forms a series of tight junctions with similar processes of adjacent Sertoli cells.
- *Blood-testis barrier* is formed by occluding junctions that unite the basal processes of adjacent Sertoli cells forming a:
  - *Basal compartment* with access to blood-borne materials and which contains spermatogonia and earliest primary spermatocytes
  - *Adluminal compartment*
    - Provides a unique microenvironment for developing germ cells that protects these cells from immunologic attack and concentrates hormones needed for sperm production
    - Contains later primary spermatocytes, secondary spermatocytes, and spermatids
- Functions
  - Mediate exchange of nutrients to germ cells
  - Form blood-testis barrier to protect developing germ cells from immunologic attack
  - Break down excess spermatid cytoplasm
  - Produce testicular fluid
  - Secrete androgen-binding protein that binds to and concentrates testosterone in the seminiferous epithelium
  - Produce inhibin, which inhibits the secretion of follicle stimulating hormone (FSH) from the adenohypophysis
  - Orchestrate movement of germ cells through germinal epithelium and facilitate cytodifferentiation and subsequent release of spermatozoa into the lumen of the seminiferous tubule

- Germ cells (spermatogenic cells)
  - Form a stratified germinal (seminiferous) epithelium
  - Cell types
    - *Spermatogonia*
      - Are diploid cells resting on the basement membrane
      - Are of two varieties. Type A spermatogonia divide mitotically to perpetuate self and to form type B cells. Type B spermatogonia divide mitotically to form primary spermatocytes.

- Undergo incomplete cytokinesis so resulting cells remain attached to each other during spermatogenesis
- Divide mitotically to produce primary spermatocytes

- *Primary spermatocytes*
  - Are the largest germ cells; each nucleus is 1.5 times larger than that of a spermatogonium.
  - Form in the basal compartment, then probably migrate through the tight junctions between Sertoli cell processes to the adluminal compartment
  - Remain in prophase about one-third of the spermatogenic cycle, so many are seen. Nuclei contain highly condensed chromosomes.
  - Are diploid cells that complete meiosis I (reductional division) to form secondary spermatocytes
- *Secondary spermatocytes*
  - Are haploid cells whose pale staining nuclei are similar in size to those of the spermatogonia nuclei
  - Are present for only eight hours of the entire 64-day spermatogenic cycle; therefore, very few are seen.
  - Divide by meiosis II (equational division) to form spermatids
- *Spermatids*
  - Are haploid cells whose nuclei are initially about two-thirds the size of spermatogonia nuclei
  - Are located near the lumen of the seminiferous tubules
  - Do not divide but undergo cytodifferentiation to form spermatozoa
    - Intercellular bridges break down.
    - Nucleus condenses and elongates.
    - Acrosome forms. An acrosome is a modified lysosome containing enzymes to aid the sperm in penetrating the zona pellucida surrounding the secondary oocyte.
    - Flagellum forms.
    - Excess cytoplasm is shed.
- *Spermatozoa*
  - Are haploid cells
  - Are anatomically mature, but incapable of fertilization at this time

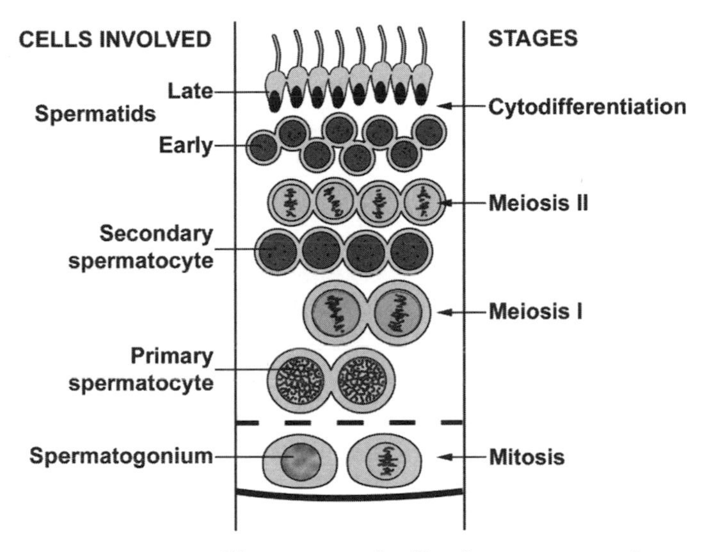

FIGURE 18.4. The stages and cells of spermatogenesis.

- Are released from Sertoli cells into the lumen of the seminiferous tubules

Spermatogenesis

- *Spermatogenesis* is defined as the process by which diploid somatic cells (spermatogonia) in the basal compartment become haploid spermatozoa lying free in the lumen of the seminiferous tubules.
- Stages
  - *Spermatocytogenesis.* Mitotic divisions of spermatogonia (diploid) to form primary spermatocytes (diploid); cytokinesis is incomplete.
  - *Meiosis.* Two cell divisions convert diploid cells to haploid (i.e., reduction of chromosomes and DNA by half); cytokinesis is incomplete.
    - Meiosis I. Primary spermatocytes (diploid) form secondary spermatocytes (haploid).
    - Meiosis II. Secondary spermatocytes form spermatids (haploid).
  - *Spermiogenesis.* Cytodifferentiation of spermatids (haploid) into spermatozoa (haploid)
  - *Spermiation.* Release of mature sperm into lumen of seminiferous tubule

- ➢ Under control of follicle stimulating hormone (FSH) from the anterior pituitary
- ➢ One cycle lasts about 64 days, with a new cycle beginning in any given location about every 16 days.

COURSE OF SPERM WITHIN THE TESTIS

- ➢ *Seminiferous tubules, convoluted portion.* Germinal epithelium where sperm production occurs; sperm are released into the lumen of this tubule from Sertoli cells.
- ➢ *Seminiferous tubules, straight portion* (tubuli recti)
  - Lined by simple columnar epithelium whose cells resemble Sertoli cells
  - Connects convoluted portion of seminiferous tubules with rete testis
- ➢ *Rete testis*
  - Is a meshwork of channels within mediastinum of testis
  - Lined by simple cuboidal cells, many of which possess a single flagellum
  - Connects the straight portion of the seminiferous tubules with efferent ducts in the epididymis

GENITAL DUCTS EXTERNAL TO THE TESTIS

EPIDIDYMIS

- ➢ The epididymis is a comma-shaped organ lying posterior to the testis that is divided into head, body, and tail subdivisions.
- ➢ *Head* region composition
  - *Efferent ducts*
    - ◆ Connect rete testis with duct of epididymis
    - ◆ Consist of about 12 ducts, each of which is coiled into a cone shape. Each duct connects with the rete testis at the apex of the cone adjacent to the testis. All ducts anastomose to form the single duct of the epididymis at the bases of the cones.
    - ◆ Form *coni vasculosi* (singular, conus vasculosus) that are composed of one coiled efferent duct plus its surrounding connective tissue, containing abundant blood vessels.
    - ◆ Are lined with a simple epithelium composed of alternating taller, ciliated cells and shorter cuboidal cells with lysosomes.

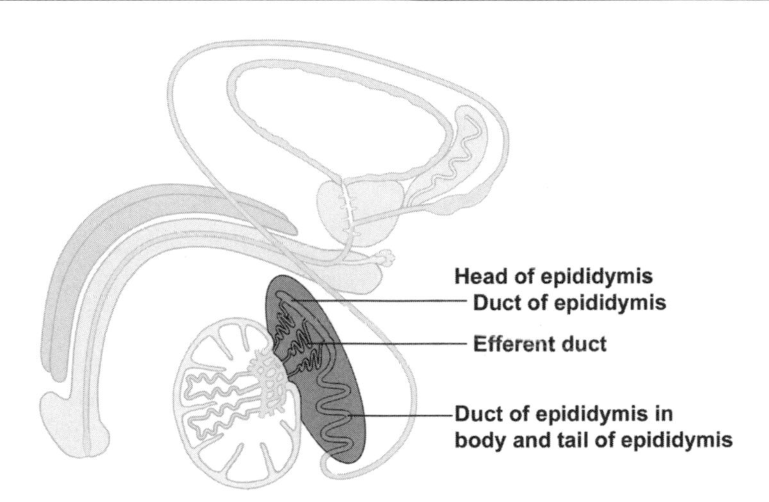

FIGURE 18.5. Components of the epididymis.

Therefore, efferent ducts present a characteristic, scalloped border adjacent to the lumen. A thin muscularis layer surrounds the epithelium.

- ◆ Function. Ciliated cells propel spermatozoa toward duct of epididymis while cuboidal cells absorb testicular fluid.

- *Duct of epididymis*. A single duct formed by fusion of efferent ducts

➢ *Body and tail regions*

- Contains the remainder of the duct of the epididymis
  - ◆ Highly coiled, single tube (6 m long) formed by union of efferent ducts in the head region
  - ◆ Lined by tall pseudostratified columnar epithelium with stereocilia, which decreases in height from head to tail regions; creates a smooth lumen when compared with efferent ducts
  - ◆ Smooth muscle layer surrounds epithelium and increases in thickness and number of layers from head to tail
- Function
  - ◆ Storage and maturation site for sperm
  - ◆ Absorption of excess testicular fluid
  - ◆ Movement of sperm toward ductus deferens

Ductus (Vas) Deferens

➢ The *ductus deferens* is a thick muscular tube carrying sperm from duct of epididymis to the ejaculatory duct.

- Structure
  - Mucosa
    - Pseudostratified columnar epithelium with stereocilia surrounds a narrow lumen.
    - Thin lamina propria
    - Longitudinal folds produce an irregular lumen.
  - Thick muscularis. Inner and outer longitudinal, middle circular layers of smooth muscle
- Course
  - Located in spermatic cord in the inguinal canal with spermatic artery, pampiniform venous plexus, and a nerve plexus
  - Enters abdominal cavity, crosses above entrance of ureter into bladder, and enlarges to form the ampulla, which lies posterior to urinary bladder
  - Is joined by duct of the seminal vesicle just before it enters the prostate
- Function. Transports sperm from epididymis to ejaculatory duct in prostate

EJACULATORY DUCT

- The ejaculatory duct is formed by the union of the ductus deferens with the duct of the seminal vesicle.
- No muscle layer is retained from the ductus deferens.
- The ejaculatory duct traverses the prostate gland to join the prostatic urethra.

URETHRA

- *Prostatic urethra.* Within prostate; lined with transitional epithelium
- *Membranous urethra.* Pierces skeletal muscle of the urogenital diaphragm; lined with stratified or pseudostratified columnar epithelium
- *Penile urethra* (discussed with penis)

GENITAL GLANDS

SEMINAL VESICLE

- *Seminal vesicles* are paired glands lying posterior to the urinary bladder.
- Each is composed of a single, highly tortuous tube.

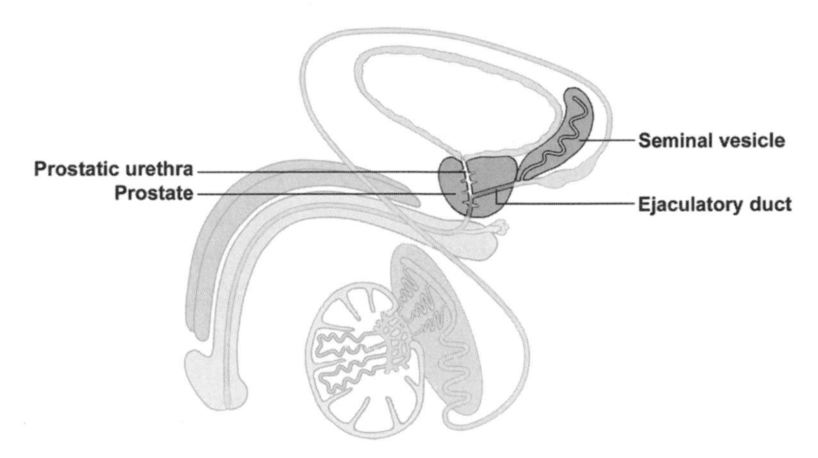

FIGURE 18.6. Major genital glands and their associated passageways.

- ➢ Function. Add sperm-activating substances, such as fructose, citrate, proteins, and prostaglandins to seminal fluid; provides bulk of seminal fluid
- ➢ Structure
  - Pseudostratified columnar epithelium with many secretory granules overlies a thin layer of connective tissue. These tissues are thrown into an intricate system of primary, secondary, and tertiary folds that produce a pattern of arcades, dividing the central lumen into fragments.
  - A thin layer of smooth muscle surrounds the tube.

PROSTATE GLAND

- ➢ The *prostate*, a single gland, is the largest of the genital glands and surrounds the prostatic urethra.
- ➢ 30–50 tubuloalveolar glands, opening onto the prostatic urethra, can be divided into groups depending on their location.
- ➢ Capsule. Dense connective tissue with abundant smooth muscle; septa from the capsule also possesses smooth muscle fibers and partition the gland into indistinct lobes.
- ➢ Usually lined by a pseudostratified columnar epithelium whose height will vary with its activity
- ➢ *Prostatic concretions*. Lamellated, spherical bodies that are the condensation of secretory products. The number of concretions increases with age.
- ➢ Function. Contributes a thin, milky fluid to semen that is rich in citric acid and acid phosphatase

PENIS

COMPOSITION

- ➢ Three cylindrical masses of erectile tissue
  - *Corpora cavernosa.* Paired dorsal cylinders
  - *Corpus spongiosum* (corpus cavernosum urethrae)
    - ◆ Single, ventral cylinder that houses the penile urethra
    - ◆ Expands to terminate in glans penis that caps the two corpora cavernosa
- ➢ Structure
  - Outer covering of skin (epidermis and dermis)
  - *Tunica albuginea*
    - ◆ Capsule of dense, nonelastic connective tissue surrounding the three cylinders
    - ◆ Thicker around corpora cavernosa than around corpus spongiosum
    - ◆ Forms an incomplete septum between the corpora cavernosa
  - Structure of erectile tissue
    - ◆ Sponge-like *cavernous spaces* (venous spaces) separated by connective tissue trabecula with smooth muscle fibers
    - ◆ Deep artery in each corpus cavernosum supplies blood to
      - ■ Nutritive arteries that supply trabecula
      - ■ *Helicine arteries* that supply cavernous spaces

PROCESS OF ERECTION

- ➢ Flaccid state is effected by a minimal blood flow to the penis. This blood flow is regulated by the continuous input of the sympathetic division of the autonomic nervous system on the tone of the smooth muscle in the penile vasculature.
- ➢ Erection
  - Parasympathetic division of autonomic nervous system effects relaxation of smooth muscle (vasodilation) of the deep and helicine arteries.
  - The subsequent filling of the cavernous spaces expands these vessels against the tunica albuginea, causing the penis to become erect and turgid.
  - Corpus spongiosum does not become as erect as the other cavernous bodies because the tunica albuginea is thinner here. Therefore, sperm can be transported during ejaculation.

- Return to flaccid state occurs with decline of parasympathetic activity.

Penile Urethra

- The *penile urethra* is located within corpus spongiosum (corpus cavernosum urethrae).
- Microscopic anatomy
  - Lined by pseudostratified columnar epithelium that becomes stratified squamous moist in fossa navicularis, the terminal enlargement in the glans penis
  - *Glands of Littre*
    - Mucus-secreting glands
    - Originate in mucus-secreting recesses of the urethra and extend obliquely toward the base of the penis
    - Secrete a mucous fluid that is the initial ejaculate; provides lubrication

Structures Identified in This Section

Testis
- Intratesticular ducts
- Testis
- Tunica albuginea
- Tunica vaginalis
- Processes vaginalis
- Tunica vaginalis, parietal layer
- Mesothelium
- Mediastinum
- Rete testis
- Seminiferous tubules, convoluted portion
- Seminiferous tubules, straight portion

Epididymis
- Body and tail
- Coni vasculosi
- Coni vasculosi, vasculature
- Duct of epididymis
- Efferent ducts
- Efferent ducts, cilia
- Efferent ducts, epithelium
- Efferent ducts, lysosomes
- Head
- Pseudostratified columnar epithelium with stereocilia
- Rete testis
- Smooth muscle
- Spermatozoa

Spermatic cord
- Ductus deferens
- Ductus deferens, epithelium
- Ductus deferens, lamina propria
- Ductus deferens, smooth muscle
- Nerves
- Pampinoform plexus of veins
- Testicular artery
- Ejaculatory duct

Penis
- Cavernous spaces
- Corpora cavernosa
- Corpus spongiosum

- Deep artery
- Glands of Littre
- Helicine artery
- Trabeculae
- Tunica albuginea
- Urethra

Seminal vesicles
- Arches
- Epithelium
- Lamina propria
- Lumen
- Smooth muscle

Prostate
- Epithelium
- Glands
- Prostatic concretions
- Smooth muscle

CHAPTER

19

EYE

GENERAL CONCEPTS

- ➢ The eyes are complex photoreceptive organs located in the bony orbits of the skull. Movement of the eye is accomplished by a set of extrinsic ocular muscles, which insert on the outer surface of the globe.
- ➢ Each eye consists of image-forming structures, a photoreceptive retina, and a fibrous globe to provide support.
- ➢ The eye is protected by an eyelid, a moveable fold of skin that covers the anterior surface of the globe.

EYELID

- ➢ Protective covering of the eye.
- ➢ Components
  - Covered on its outer surface by thin skin; possesses hair follicles, eyelashes, sebaceous glands, and sweat glands
  - *Tarsal plate.* Region of dense fibrous and elastic connective tissues within the eyelid that provide support
  - *Meibomian glands.* Specialized sebaceous glands on the inner surface of the eyelid whose secretions add to the tear film to reduce evaporation

Digital Histology: An Interactive CD Atlas with Review Text, by Alice S. Pakurar and John W. Bigbee
ISBN 0-471-64982-1

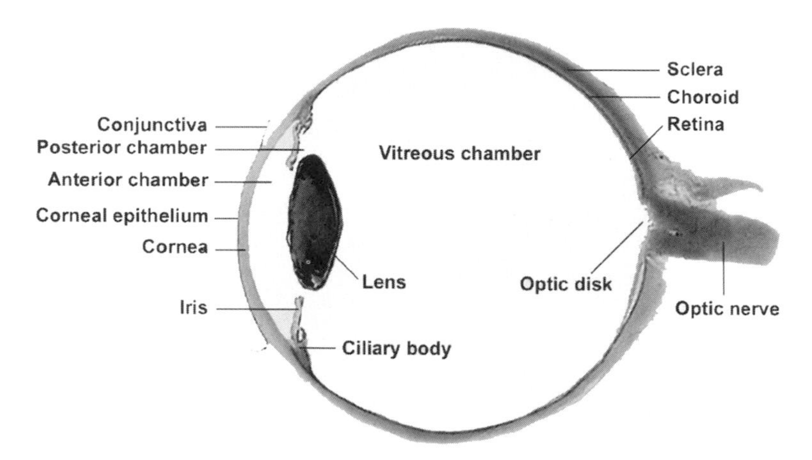

FIGURE 19.1. Midsagittal section of the eyeball.

- Contains the *obicularis oculi muscle*
- *Conjunctiva.* Lines the inner surface, consisting of a stratified columnar epithelium with goblet cells; the conjunctiva is reflected onto the globe as the bulbar conjunctiva, which is continuous with the *corneal epithelium.*

EYEBALL (GLOBE)

- Composed of three layers or tunics
  - *Fibrous tunic* consisting of the sclera and cornea
  - *Vascular tunic* or *uveal tract* consisting of the iris, ciliary body, and choroid
  - *Neural tunic* consisting of the retina
- Contains three chambers
  - *Anterior chamber* is the space between the cornea and the iris, filled with *aqueous humor* fluid.
  - *Posterior chamber* lies between the iris anteriorly and the lens, ciliary body, and zonule fibers posteriorly; filled with aqueous humor
  - *Vitreous chamber* is located behind the lens and is filled with a gelatinous substance called the *vitreous body.*

FIBROUS TUNIC OF THE EYE—OUTER TUNIC

- *Sclera*
  - Opaque layer composed of dense, irregular connective tissue; forms the outer layer of the posterior four-fifths of the globe

- Gives shape and support for the globe
- Provides insertion points for extraocular muscles

➢ *Cornea*

- Anterior continuation of the sclera, covering the anterior one-fifth of the eye
- Transparent and avascular; transparency results from the ordered arrangement of its collagen fibers and low state of tissue hydration.
- Convex curvature aids in focusing light (refraction).
- Layers (anterior to posterior)
  - *Corneal epithelium.* Covers the anterior surface of the cornea; composed of a moist, stratified squamous epithelium that is continuous with the bulbar conjunctiva at the *limbus*
  - *Bowman's membrane.* Acellular collagenous layer beneath the corneal epithelium
  - *Stroma.* Multiple layers of parallel collagen fibers constitute the majority of the cornea. The collagen fibers in each layer are arranged at about right angles to adjacent layers. The highly ordered arrangement of these fibers contributes to the transparency of the cornea.
  - *Descemet's membrane.* Thickened basal lamina of the corneal endothelium
  - *Corneal endothelium.* Simple squamous epithelium covering the posterior surface of the cornea; regulates the hydration state of the stroma

➢ *Corneo-scleral junction (limbus)*

- Transition zone between the cornea and the sclera
- Bowman's membrane ends and the corneal epithelium thickens at this junction.
- *Trabecular meshwork.* Irregular channels in the stroma that are lined by endothelium. Drains the aqueous humor from the anterior chamber to maintain proper intraocular pressure. The channels of the trabecular meshwork merge to form the *canal of Schlemm*, a ring-like sinus that encircles the limbus and drains into the venous system.

VASCULAR TUNIC (UVEAL TRACT) OF THE EYE—MIDDLE TUNIC

➢ *Choroid*

- Highly vascular, cellular layer lying beneath the sclera; this layer is richly pigmented due to the large numbers of *melanocytes*. Its inner portion is the *choriocapillary layer*, which contains large numbers of small vessels and capillaries and serves a nutritive function for the retina.

- *Bruch's membrane.* A thin layer separating the retina from the choriocapillary layer; represents the combined basal laminae of the capillary endothelium and the pigment epithelium of the retina and an intervening network of elastic and collagen fibers

➢ *Ciliary body*

- Anterior expansion of the choroid forming a ring that encircles the lens; appears triangular in cross-section
- Composed of a core of connective tissue and muscle; lined on its vitreal surface by two layers of columnar cells, an inner pigmented epithelium and an outer layer of nonpigmented cells. This layer, the nonsensory retina, represents the attenuated anterior part of the sensory layer of the retina.
- *Ciliary processes*
    - ◆ Ridge-like extensions from the ciliary body
    - ◆ *Zonule fibers.* Emerge from between the processes and attach to the lens capsule
    - ◆ The aqueous humor is produced by the epithelium of the ciliary processes.
- *Ciliary muscles.* Smooth muscle fibers that insert on the sclera and ciliary body; contraction of circularly arranged fibers releases tension on the zonule fibers, allowing the lens to assume a more spherical shape, thus providing for focusing on near objects (accommodation). Contraction of radially oriented smooth muscle fibers results in flattening of the lens, thus providing for focusing on far objects.

➢ *Iris*

- Disc-shaped structure that arises from the anterior margin of the ciliary body; separates anterior and posterior chambers and partially covers the lens
- Composed of loose connective tissue that is covered on its anterior surface by an incomplete layer of pigment cells and fibroblasts. Its posterior surface is covered by a double layer of pigmented epithelial cells.
- *Pupil.* Central opening in the iris, the diameter is regulated by contraction of two sets of intrinsic smooth muscle in the iris.
    - ◆ *Dilator pupillae muscle.* Derived from the more anterior, pigmented epithelial layer; consists of radially oriented cells whose contraction widens the aperture of the pupil

- *Constrictor pupillae muscle.* Consists of circularly oriented smooth muscle fibers surrounding the pupil; contraction of these fibers decreases the diameter of the pupil.

RETINA—INNER TUNIC

- Inner-most of the three layers, forming a cup-shaped structure. The posterior portion is photosensitive and extends forward to the ciliary body, terminating as the *ora serrata*. The nonphotosensitive anterior portion is reduced in thickness and number of layers and forms the posterior lining of the ciliary body and the posterior lining of the iris.
- The photosensitive portion contains the photoreceptors, which transduce light into nervous impulses, and neurons, which perform the initial integration of the visual signals.
- Overview of retinal cytoarchitecture
  - Basic plan of the *retina* consists of a three cell pathway
    - *Rods* and *cones*. Photoreceptors that transduce light energy into neural activity and form the *photoreceptor layer*; their nuclei are located in the *outer nuclear layer*.
    - *Bipolar cells*. Synapse with rods and cones; nuclei are located in the *inner nuclear layer*.
    - *Ganglion cells*. Synapse with bipolar cells; cell bodies are located in the *ganglion cell layer*; axons from these cells form the *optic nerve fiber layer* as they pass toward the *optic disc*, head of the optic nerve.
  - Regions of synaptic integration
    - *Outer plexiform layer.* Location of synapses of rods and cones with bipolar cells
    - *Inner plexiform layer.* Location of synapses of bipolar cells and ganglion cells
- Layers of the retina–from outer to inner
  - Composed of 10 layers. The naming of the layers is based on their position relative to the path of the neural conduction, not the path of light.
  - *Pigment epithelium*
    - Cytoplasm contains numerous melanin granules to absorb light and reduce reflection
    - Columnar epithelial cells with apical microvilli whose bases are adherent to Bruch's membrane in the choroid

    - Cells posses a cylindrical sheath that surrounds the apical tips of the photoreceptors; these sheaths aid in phagocytosis and digestion of membranous discs shed by the photoreceptors.
- *Photoreceptor layer*
    - Composed of *rods* and *cones*
    - Rods are sensitive to low light intensity, outnumber cones and are located throughout the retina
    - Cones are less numerous than rods, sensitive to high intensity light and respond to color. Cones provide greater visual acuity and are concentrated in the *fovea centralis*. (See below.)
    - *Outer segment.* Contains flattened, membranous discs that contain the visual pigments rhodopsin (rods) and iodopsins (cones).
    - *Inner segment.* Separated from the outer segment by a constriction, contains the major synthetic and energy-producing organelles.
- *External limiting membrane.* Not a true membrane; formed by adherent junctions of Mueller cells, modified astrocytes, with the photoreceptors
- *Outer nuclear layer.* Location of the nuclei of rods and cones
- *Outer plexiform layer.* Region of synaptic contacts between photoreceptor axons and bipolar cell dendrites
- *Inner nuclear layer.* Location of cell bodies of bipolar cells. Also present are additional neurons, amacrine and horizontal cells.
- *Inner plexiform layer.* Location of synaptic contacts between bipolar cell axons and ganglion cell dendrites.
- *Ganglion cell layer.* Location of cell bodies of ganglion cells
- *Optic nerve fiber layer.* Collections of unmyelinated ganglion cell axons that pass toward the *optic disc*, the head of the optic nerve, where they exit to form the optic nerve (cranial nerve II).
- *Internal limiting membrane.* Formed by the basal portions of Mueller cells

➢ *Fovea centralis.* Region of the retina providing greatest visual acuity, consists entirely of cones; other retinal layers are displaced centrifugally to allow for an unimpeded path for the light to reach the photoreceptors.

➢ *Optic disc* ("blind spot"). Region composed only of axons from retinal ganglion cells as they pass through the sclera to form the optic nerve

Lens

- Biconcave, transparent, and elastic
- Suspended by radially oriented zonule fibers that extend from the ciliay body to insert into the lens capsule
- Structure of the lens
  - *Lens capsule.* A thickened basal lamina, produced by the subcapsular epithelium, surrounds the entire lens.
  - *Subcapsular epithelium.* Simple cuboidal epithelium, present only on the anterior surface of the lens; apical surfaces of the cells are directed toward the center of the lens.
  - *Lens fibers.* Derived from cells of the subcapsular epithelium primarily in the equatorial region of the lens; lens fibers are highly differentiated cells that lose their organelles and become filled with crystallin proteins.
- Contraction of the ciliary muscle releases tension on the zonule fibers, allowing the lens to assume a more spherical shape which provides for focusing on near objects (accommodation).

Structures Identified in This Section

Chambers
- Anterior chamber
- Posterior chamber
- Vitreous chamber

Eyelid
- Conjunctiva
- Eyelash
- Hair follicles
- Meibomian glands
- Skin
- Tarsus

Fibrous tunic
- Bowman's membrane
- Canal of Schlemm
- Cornea
- Corneal endothelium
- Corneal epithelium
- Corneo-scleral junction (limbus)
- Descemet's membrane
- Sclera
- Stroma
- Trabecular meshwork

Lens
- Lens capsule
- Lens fibers
- Subcapsular epithelium
- Neural tunic
- External limiting membrane
- Fovea centralis
- Ganglion cell layer
- Inner nuclear layer
- Inner plexiform layer
- Internal limiting membrane
- Optic disc
- Optic nerve
- Optic nerve fiber layer
- Ora serrata
- Outer nuclear layer
- Outer plexiform layer
- Photoreceptor layer containing rods and cones

Pigmented epithelium

Vascular tunic (uveal tract)
- Bruch's membrane
- Chorio-capillary layer
- Choroid
- Ciliary body
- Ciliary muscle
- Ciliary processes
- Constrictor pupillae muscle
- Dilatory pupillae muscle
- Iris
- Melanocytes
- Pupil

CHAPTER

20

EAR

COMPONENTS

- ➢ *External ear.* Receives sound waves, transmitting them to the tympanic membrane
- ➢ *Middle ear.* Transmits movement of the tympanic membrane by three ear ossicles to fluid in the inner ear
- ➢ *Inner ear.* Contains a receptor that responds to these fluid vibrations for the perception of sound. Additional receptors in the inner ear respond to the effects of gravity and motion of the head to maintain equilibrium.

EXTERNAL EAR

- ➢ *Auricle* or *pinna.* Shallow appendage on the lateral surfaces of the head that is formed by thin skin covering a framework of elastic cartilage
- ➢ *External auditory meatus.* Short tube leading to the tympanic membrane
  - The thin skin, lining the meatus, possesses ceruminous glands. Their secretions combine with those of adjacent sebaceous glands to form cerumen, a thick, waxy product.

Digital Histology: An Interactive CD Atlas with Review Text, by Alice S. Pakurar and John W. Bigbee
ISBN 0-471-64982-1

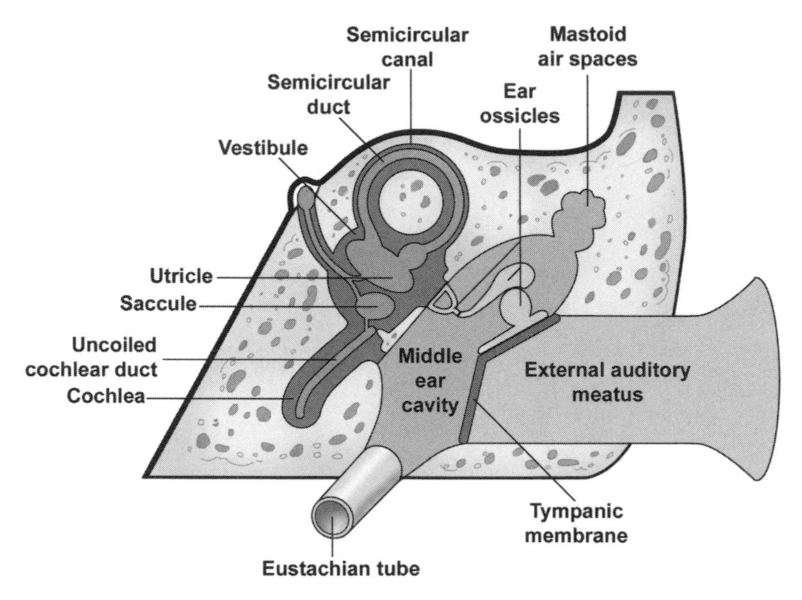

FIGURE 20.1. Schematic illustration of the three subdivisions of the ear.

- Support provided by
  - ◆ Elastic cartilage in the outer portion
  - ◆ Temporal bone in the inner portion

➢ *Tympanic membrane* (ear drum) separates external from the middle ear.

- Composition (from exterior to interior). Thin skin, two layers of collagen and elastic fibers with radial then circular arrangements, and a mucous membrane that is continuous with that lining the middle ear
- Attachment of the malleus, an ear ossicle, to the inner surface pulls the tympanic membrane into a flattened, cone shape.

MIDDLE EAR (TYMPANIC CAVITY)

➢ The *middle ear* or *tympanic cavity* is a cavity within the temporal bone that is bounded by the tympanic membrane laterally and the bony wall of the inner ear medially. It communicates with the mastoid air cells posteriorly, and with the nasopharynx, via the Eustachian tube, anteriorly.

➢ Structure

- Lined by a mucous membrane whose epithelium is predominately simple squamous

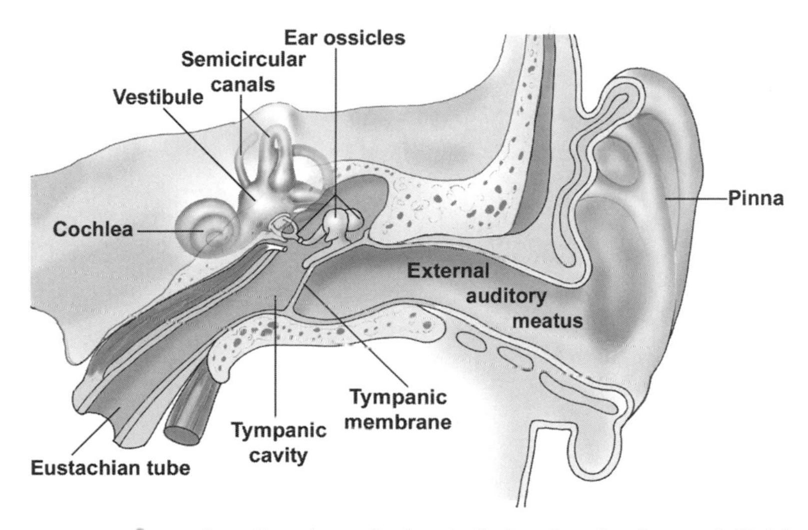

FIGURE 20.2. Coronal section through the skull showing the three subdivisions of the ear in the temporal bone.

- Ear ossicles, small bones, transmit vibrations from the tympanic membrane to the inner ear.
  - Components
    - *Malleus.* Attached to the tympanic membrane
    - *Incus.* Interconnects malleus with stapes
    - *Stapes.* Footplate of the stapes fits into the oval window of the inner ear
  - Ossicles are connected to each other by ligaments and are covered with mucosa.
  - Small muscles attached to malleus (tensor tympani) and stapes (stapedius) modulate vibrations of these ossicles.
- *Eustachian tube* (auditory tube)
  - Connects middle ear with the nasopharynx
  - Is lined by a mucous membrane whose epithelium becomes pseudostratified near the nasopharynx. Cilia associated with this epithelium beat toward the pharynx.
  - Is supported first by bone and then by cartilage and fibrous tissue as it nears the nasopharynx
  - Is usually collapsed but opens during swallowing to equilibrate air pressure

- *Oval window* and *round window*
  - Openings in the petrous portion of the temporal bone that form the medial wall of the middle ear
  - The oval window is occupied by the footplate of the stapes.
  - The round window is covered by a membrane that bulges to relieve pressure in the cochlea that originates from motion of the stapes at the oval window.
- *Mastoid air spaces*, located in the mastoid process of the temporal bone, communicate posteriorly with the middle ear.

INNER EAR

- The *inner ear* is located in the *petrous portion* of the *temporal bone.*
- Components
  - *Osseous labyrinth.* Series of interconnected tubular and cavernous spaces in the petrous portion of the temporal bone that are lined with periosteum and filled with perilymph fluid
    - *Vestibule.* Centrally located chamber; communicates with middle ear via the oval window
    - *Semicircular canals*
      - Are three tubular spaces that communicate with and lie posterolaterally to the vestibule
      - Are oriented in three mutually perpendicular planes
      - An enlargement at one end of each canal, adjacent to the vestibule, houses the ampulla of the semicircular ducts.
    - *Cochlea.* An osseous tube that connects with and lies anteromedially to the vestibule
      - Tube is coiled into a spiral shape with 2.5 turns, resembling a snail shell.
      - The tube's spiraling in the temporal bone results in the formation of a central, bony axis for the cochlea called the *modiolus,* which resembles a screw. The threads of the screw project into the cochlea and are called the *osseous spiral lamina.*
      - The modiolus houses the cochlear division of *cranial nerve VIII* and its sensory ganglion, the *spiral ganglion.*
  - *Membranous labyrinth.* Series of interconnected ducts and chambers that are suspended within the osseous labyrinth. Contain the fluid, endolymph. These ducts and chambers contain receptors for hearing and for static and kinetic senses.

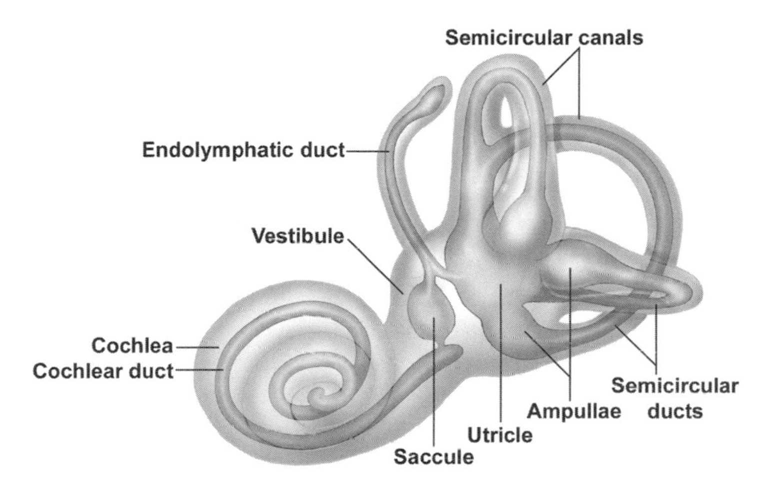

FIGURE 20.3. Inner ear: the membranous labyrinth is suspended in the osseous labyrinth.

- *Utricle* and *saccule*. Suspended within the vestibule. A receptor, the *macula*, in each of these two chambers responds to stimuli of linear acceleration and gravitational forces.
- *Semicircular ducts* (three). One duct is suspended in each of the semicircular canals; both ends of each duct connect with the utricle. An enlargement, the *ampulla*, at one end of each duct is located in the enlargement of each semicircular canal and contains a receptor, the *crista ampullaris*, for angular acceleration.
- *Cochlear duct*. Located in the center of the cochlea. The cochlear duct communicates indirectly with the saccule. The receptor in the cochlear duct, the *organ of Corti*, responds to sound vibrations.
- *Endolymphatic duct*. Formed by union of small ducts from the utricle and saccule; extends toward the brain where it terminates as an enlargement, the endolymphatic sac, between layers of the meninges. Probably functions to absorb *endolymph*.

- Sensory innervation is provided by *cranial nerve VIII*, the vestibulocochlear nerve.

➢ *Utricle* and *saccule*

- Portions of the membranous labyrinth that are connected to each other and are suspended in the osseous vestibule
- *Macula*. Receptor in both the utricle and saccule
  - Thickening in the wall of the utricle and saccule composed of:

- *Supporting cells*
- *Hair cells* with stereocilia and a cilium (kinocilium) that are embedded in the gelatinous layer
- *Gelatinous layer* is produced by supporting cells and covers both these and the hair cells.
- *Otoliths* (otoconia). Calcium carbonate crystals that are suspended at the top of the gel

◆ Linear acceleration and the force of gravity displace the otoliths, stimulating the stereocilia and kinocilia and initiating a neural, sensory impulse in the vestibular division of cranial nerve VIII.

➢ *Semicircular ducts* (three)

- Portions of the membranous labyrinth suspended in the osseous semicircular canals; both ends of each semicircular duct connect to the utricle.
- *Crista ampullaris*. Receptor in the ampullary enlargement of each semicircular duct
  - ◆ Ridge-like structure that lies perpendicular to the long axis of each duct. Internal cell structure is similar to that of a macula except:
    - Gelatinous layer, called the *cupula*, is shaped like a cone and extends across the ampulla to the opposite wall, thus spanning the duct.
    - Otoliths are absent.
  - ◆ Angular acceleration displaces the cupula that deflects the stereocilia and kinocilia and initiates a neural, sensory impulse in the vestibular division of cranial nerve VIII.
  - ◆ Orientation in three distinct planes allows for complex detection of motion.

➢ *Cochlear duct*

- Wedge-shaped duct of the membranous labyrinth suspended in the middle of the tubular, osseous cochlea. Position of the cochlear duct separates the bony cochlea into three subdivisions.
  - ◆ *Scala vestibuli*. This subdivision of the cochlea is continuous with the vestibule and lies above the cochlear duct, separated from it by the vestibular membrane.
  - ◆ *Cochlear duct*. Contains the receptor for sound. The cochlear duct is located in the middle of the cochlea and is continuous with the saccule through a small duct. Its roof is the vestibular membrane separating it from the osseous scala vestibuli. Its floor is formed by the basilar membrane that is continuous with the

osseous spiral lamina; both separate the cochlear duct from the scala tympani.

- *Scala tympani.* Subdivision of the bony cochlea lying beneath the cochlear duct. The scala tympani is continuous with the scala vestibuli at the *helicotrema,* located at the tip of the cochlea. The scala tympani terminates at the round window where pressure on the perilymph in this scala, initiated at the oval window and transported through scala vestibuli to scala tympani, is released.

- *Organ of Corti.* Receptor for sound in the cochlear duct; positioned on the floor of the cochlear duct, resting on the basilar membrane
  - Structure
    - *Supporting cells.* Several varieties, including pillar cells that form the boundary of a triangular space called the inner tunnel. Provide support for the hair cells, among other functions.
    - *Inner and outer hair cells.* Receptor cells located on either side of the inner tunnel possess stereocilia that are embedded in the tectorial membrane.
    - *Tectorial membrane.* This gelatinous membrane extends over the hair cells. Stereocilia of the hair cells are embedded in the tectorial membrane.
  - Discrimination of sound

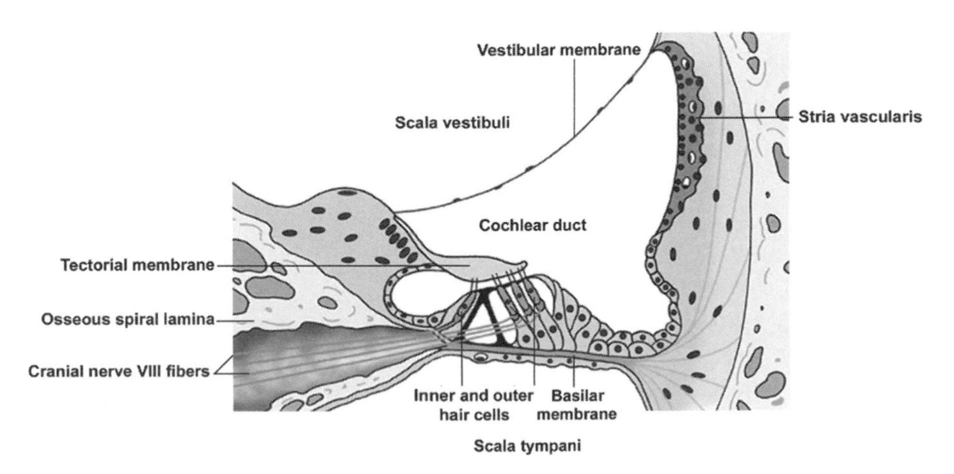

FIGURE 20.4. Cochlear duct, the receptor for sound, is a part of the membranous labyrinth in the inner ear.

- Inward movement of the stapes at the oval window generates pressure on the perilymph in the vestibule that is transmitted into the scala vestibuli.
- From the scala vestibuli, pressure is conducted, by deflection of the vestibular membrane, to the endolymph of the cochlear duct and to the basilar membrane. Movement of the basilar membrane into scala tympani and away from the tectorial membrane causes a shearing force on the stereocilia embedded in this membrane and initiates a neural, sensory response in the cochlear division of cranial nerve VIII.
- Sound vibrations in the scala vestibuli also continue into the scala tympani at their junction at the helicotrema.
- Sound vibrations in scala tympani are relieved by the bulging of the round window into the middle ear.

- *Stria vascularis* is a vascularized epithelium located on the outer wall of the cochlear duct that produces endolymph.

STRUCTURES IDENTIFIED IN THIS SECTION

- External ear
  - Ceruminous glands
  - Elastic cartilage
  - Hair follicles
  - Lumen
  - Perichondrium
  - Sebaceous glands
  - Skeletal muscle
  - Sweat glands
  - Thin skin
  - Tympanic membrane
- Middle ear
  - Auditory meatus
  - Auditory ossicles
  - Cochlea
  - Eustachian tube
  - Mastoid air cells
  - Mucosa
  - Mucosal epithelium
  - Osseous labyrinth
  - Stapedius muscle
  - Tensor tympani muscle
  - Tympanic membrane
  - Window, oval
  - Window, round
- Inner Ear
  - Bipolar neurons
  - Cranial nerve VIII
  - Osseous labyrinth
    - Cochlea
      - Modiolus
      - Osseous spiral lamina
    - Helicotrema
    - Scala tympani
    - Scala vestibuli
    - Semicircular canals
      - Ampulla
    - Vestibule
  - Membranous labyrinth
    - Cochlear duct
      - Basilar membrane
      - Organ of Corti
        - Hair cells
        - Spiral limbus
        - Stria vascularis
        - Tectorial membrane

- Vestibular membrane
- Endolymphatic duct
- Saccule
  - Macula
- Semicircular ducts
  - Ampullae
  - Crista ampullaris
  - Connective tissue reticulum
  - Cupula of crista
  - Hair and supporting cells
  - Planum semilunatum
- Utricle
  - Macula
    - Otoliths
    - Planum semilunatum
- Peripheral axons
- Spiral ganglion
- Temporal bone
- Window, oval
- Window, round

INDEX

Digital Histology: An Interactive CD Atlas with Review Text, by Alice S. Pakurar and John W. Bigbee
ISBN 0-471-64982-1

Digital Histology Version 4.0

The main program should run automatically when you insert the disc into your CD-ROM drive. If it does not, you can start the program by navigating to the CD-ROM drive and double-clicking on the histology.exe icon (HistologyMac on the Macintosh).

Virtual Slide

Digital Histology Version 4 includes a virtual slide of a kidney section that allows you to zoom from low to high magnification and pan in all directions in the plane of section. Some users have had difficulty with the virtual slide using browsers other than Internet Explorer. Unless you are an advanced user, you should use Internet Explorer. Detailed instructions are included in the program on the CD-ROM.

Program Performance

If your computer is several years old, you may find the program runs rather slowly. You can improve the speed by copying the contents of the CD to your computer hard drive:

1. Create a folder on your desktop.
2. Browse the contents of the CD, as follows:
   Windows—Click with the right mouse button on the Start button.
     Select Explore from the pop-up menu.
     Browse through the Windows Explorer window until you find the DigHisto4 CD icon.
     Click once to display the contents of the CD.
   Macintosh—Find the DigHisto4 CD icon on your desktop.
     Double-click to open the CD and display its contents.
3. Select all of the contents. Drag the selected contents to the new folder you created on your desktop.
4. When the computer has finished transferring the files, open the folder on your desktop containing the program files. Double-click on histology.exe (HistologyMac on the Macintosh) to start the main program.

Display Properties

Digital Histology is designed to display at a screen resolution of 640 × 480 pixels. If your monitor is set to display a higher resolution, the program screen will appear relatively small. If you wish to display it at full size, set your display properties to 640 × 480:

Windows

Close or minimize all running applications and click with the right mouse button on your desktop. This will bring up a pop-up window. Select Display Properties.

Click on the Settings tab. You will see a panel labeled "Screen resolution" or "Screen area." Inside that panel is a slider. Move the slider until the resolution reads 640 × 480.

Click OK. The screen will go blank, and an alert box will pop up asking you to confirm your decision to change the screen resolution. Click "Yes" or "OK" to keep these settings.

You can revert to your old display properties settings later by following the same procedure to set the resolution higher.

Macintosh

From the Apple menu, select Control Panels, then Monitors (or Monitors and Sound).

In the Resoultion panel, click on 640 × 480. Your screen resolution will change immediately.

Close the control panel.

You can revert to your old display properties settings later by following the same procedure to set the resolution higher.

System Requirements

Windows

- Pentium or higher processor
- Windows 95, 98, 2000, Me, XP, or NT4
- 64MB of RAM
- Monitor capable of at least 640 × 480 screen resolution and at least a 16-bit color CD-ROM drive
- Internet Explorer or Netscape Version 4 or later—it is strongly recommended that you set up Internet Explorer as your default browser for viewing the virtual slide.

Macintosh

- PowerPC processor (G3 or better recommended)
- MacOS 8 or 9—Digital Histology is not OSX compatible, and does not work reliably in Classic mode. If you use OSX, you will need to change the Startup Disk in your System Preferences to reboot in MacOS 9.
- 64MB of RAM
- Monitor capable of at least 640 × 480 screen resolution and at least a 16-bit color CD-ROM drive
- Internet Explorer or Netscape Version 4 or later—it is strongly recommended that you set up Internet Explorer as your default browser for viewing the virtual slide.